Recent Advances in Mineralogy

Edited by Miloš René

Published in London, United Kingdom

Recent Advances in Mineralogy
http://dx.doi.org/10.5772/intechopen.107567
Edited by Miloš René

Contributors
Miloš René, Zdeněk Dolníček, Khalihena Groune A.L., Jun-Ichi Yoshimura, Luqman Kareem Salati, Moses Shola Adeyemo, Yohichi Kohzuki, Young-Seog Kim, YoungJae Kim, Francois Hategekimana, Mohammed S. M. Adam, Ying Li, Rui Wu

First published in London, United Kingdom, 2023 by IntechOpen
IntechOpen is the global imprint of INTECHOPEN LIMITED, registered in England and Wales, registration number: 11086078, 5 Princes Gate Court, London, SW7 2QJ, United Kingdom

British Library Cataloguing-in-Publication Data
A catalogue record for this book is available from the British Library

Additional hard and PDF copies can be obtained from orders@intechopen.com

Recent Advances in Mineralogy
Edited by Miloš René
p. cm.
Print ISBN 978-1-83768-978-1
Online ISBN 978-1-83768-979-8
eBook (PDF) ISBN 978-1-83768-980-4

For EU product safety concerns:
IN TECH d.o.o., Prolaz Marije Krucifikse Kozulić 3, 51000 Rijeka, Croatia,
info@intechopen.com or visit our website at intechopen.com.

Meet the editor

Dr. Miloš René, Ph.D., is a scientist at the Institute of Rock Structure and Mechanics, Academy of Sciences of the Czech Republic. He obtained his MSc in Mineralogy and his Ph.D. in Economic Geology from Charles University, Prague, Czech Republic in 1968 and 1981, respectively. He has published about eighty-five scientific papers in peer-reviewed journals in the fields of mineralogy, petrology, geochemistry, and economic geology. His recent scientific activity is concentrated on petrology and geochemistry of granitic rocks, mineralogy of accessory minerals, and mineralogy of uranium deposits, especially from the Central European Variscan belt.

Contents

Preface

Mineralogy is a branch of geosciences that deals with crystal structures, physical properties, and chemical composition of minerals in various igneous, metamorphic, and sedimentary rocks and processes, from high to low temperatures and in various parts of Earth.

Mineralogy is also an important background for some other geosciences and of significance for many other scientific and technical disciplines, including the mining and materials industry.

Recent Advances in Mineralogy includes nine chapters that discuss the mineralogy, geochemistry, and mechanical properties of different granitic rocks, the origin and properties of different synthetic minerals, and the mineralogy of bituminous rocks from North Africa.

Chapter 1, "Introductory Chapter: Mineralogy, Geochemistry and Metallogeny of Granites" by Miloš René, focuses on the main problems and solutions of mineralogy, geochemistry, and metallogeny of granitic rocks.

Chapter 2, "Granitoids of the Mauthausen Type in the Czech Part of the Moldanubian Batholith" by Miloš René and Zdeněk Dolníček, examines geological setting, petrology, geochemistry, and association of accessory minerals in biotite granodiorites of the Mauthausen type, which represent the youngest group of granitic rocks in the Moldanubian Batholith.

Chapter 3, "Fluorocarbonates from Biotite Granodiorite, Slavkovský les Mts., Czech Republic" by Miloš René and Zdeněk Dolníček, examines the mineralogy of parasite, bastnäsite, and synchysite evolved in granitic rocks of the redwitzite suite from the western part of the Bohemian Massif.

Chapter 4, "Deciphering Magmatic Evolution through Zoned Magmatic Enclaves and Composite Dikes: An Example from the Late Cretaceous Taejongdae Granite in Busan, Korea" by Mohammed S.M. Adam, Francois Hategekimana, YoungJae Kim, and Young-Seong Kim focuses on petrology, mineral chemistry, age dating, and geochemistry of Taejongdae granite in Southeastern Korea.

Chapter 5, "Examining the Effect of Powder Factor Variability on Granite Productivity" by L. K. Salati and M. S. Adeyemo, discusses granite productivity during its quarrying.

Chapter 6, "Physical and Mechanical Properties of Herrnholz Granite: An Ideal Experimental Material" by Ying Li and Rui Wu, focuses on the physical and mechanical properties of Herrnholz granite, which is found in the Hauzenberger Pluton in the southwest part of the Bohemian Massif.

Chapter 7, “Production and Applications of Synthetic Quartz” by Jun-Ichi Yoshimura, discusses the production of synthetic quartz and its applications in industry.

Chapter 8, “Dislocation-Point Defects-Induced by X-Irradiation Interaction in Alkali Halide Crystals” by Yohichi Kohzuki, reviews dislocation defects induced in KBr crystals by X-ray irradiation.

Finally, Chapter 9, “Mineralogical Properties of Moroccan Rif bituminous Rocks” by Khalihena Groune A.L., highlights the detailed geochemistry and mineralogy of bituminous rocks in the Moroccan Rif, North Africa.

Miloš René
Institute of Rock Structure and Mechanics, v.v.i.,
Czech Academy of Sciences,
Prague, Czech Republic

Section 1

Granitic Rocks

Chapter 1

Introductory Chapter: Mineralogy, Geochemistry and Metallogeny of Granites

Miloš René

1. Introduction

Granites are usually a medium- to coarse-grained magmatic rocks composed mostly of quartz, K-feldspar and plagioclase. The granites originated from magma with a higher content of silica and alkali metal oxides, which cool slowly and solidify in different levels of the continental crust. Granites in this continental crust form bodies of highly different size and form. The biggest bodies of granites form the batholiths, which could expose over more as hundreds of square kilometres. Granites are the most significant member of a larger family of granitic rocks and/or granitoids. The individual members of granitic rocks are classified according to their content of quartz, K-feldspar and plagioclase, based on international QAPF classification that was adopted by the International Union of Geological Sciences (IUGS) [1]. The granitic rocks and/or granites usually contain as minor mineral components such as micas (biotite, muscovite, Li-micas), amphiboles and pyroxenes. For granites, granitic rocks are also significant content of accessory minerals, as well as ilmenite, magnetite, apatite, zircon, monazite and xenotime. In some types of granites, which are associated with Sn-W mineralisation, are very significant accessory mineral topaz. Granites have significant highly different textures, from equigranular to porphyritic. Some cases have individual mineral grains, mostly grains of K-feldspars larger than the rock groundmass that form phenocrysts of different size (usually up to 2–5 centimetres) (**Figure 1**). Granites that occur in nature are usually lighter-coloured rocks with scattered darker minerals such as biotite, amphibole and/or relative scarce pyroxene.

2. Mineralogy

Mineralogy of granites was all the time studied using polarisation microscopy. For this research are recently used highly sophisticated polarisation microscopes produced by the Carl Zeiss and Leica Microsystems companies from Germany and/or the Olympus company from Japan. All these polarisation microscopes use different number of oculars and objectives. For presentation of microphotographs, microscopes use highly sophisticated digital cameras having high resolution. For simple manipulation with these cameras, good software is provided, which could also

Figure 1.
Phenocryst of zoned K-feldspar in two-mica granite, Smrčiny pluton, Bohemian massif.

analyse these pictures with different possibilities of qualitative and/or quantitative analyses of investigated granite samples.

Some other, also frequently used microscopic investigations of granites are using cathodoluminescence. Cathodoluminescence imaging is a very powerful tool to identify the intergrowth textures of different minerals, especially of quartz, K-feldspar and topaz (**Figures 2** and **3**). For the cathodoluminescence in mineralogy are used two different methods. By using of optical microscope cathodoluminescence (OM-CL), the CL spectra and colour images are obtained using a hot-cathode luminescence microscope. For the study of electron microscope cathodoluminescence (SEM-CL) are used electron microscopes and or electron microprobes equipped by different CL detectors [2–4]. The SEM-CL cathodoluminescence produced more detailed pictures of intergrowth textures, used also for detailed study of zoning of accessory minerals (zircon). For detailed microscopic study, different accessory minerals, especially zircon, are recently used through scanning electron microscopes.

3. Geochemistry

The chemical analyses of granites and their rock-forming and accessory minerals are recently very useful methods, which are sometimes distinctly more used in comparison with microscopy-oriented investigations of these rocks. For chemical analysis, main rock-forming components are recently used, predominantly the X-ray-fluorescence analysis (XRF), without decomposition of rock samples in solutions. For the analysis of rare elements, as well as for instance Ba, Sr., Rb, Zr, Y and REE

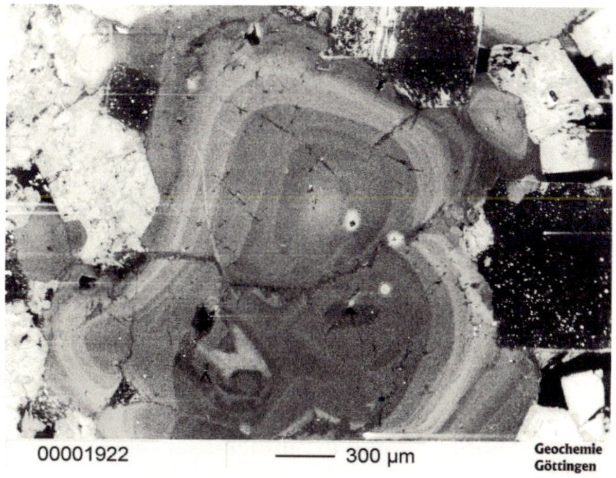

Figure 2.
Growth structures of late-magmatic snowball quartz from highly fractionated lithium granite, Krásno-Horní Slavkov Sn-W ore deposit, Bohemian massif.

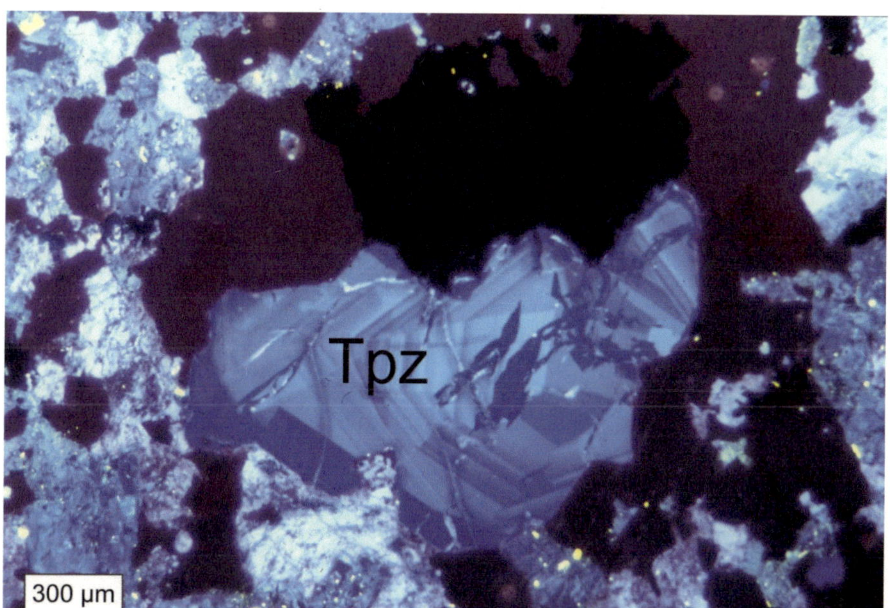

Figure 3.
Growth structures of topaz from highly fractionated lithium granite, Krásno-Horní Slavkov Sn-W ore deposit, Bohemian massif (Tpz—Topaz).

are recently used two different analytical methods, instrumental neutron activation analysis (INAA) and/or inductive coupled plasma emission mass spectrometry (ICP-MS) [5]. The XRF analyses are produced usually by the majority of mineralogical and geochemical institutes and/or geological surveys in different countries in the whole world. The production of INNA analysis is more complicated and is coupled with the presence of nuclear reactors. The production of ICP-MS analysis is concentrated on relatively a small number of geochemical institutes. However, the INNA and ICP-MS analyses of different geological materials could be well obtained from two geochemical laboratories in Canada (Act Labs and ACME). Both commercial laboratories are equipped with very sophisticated laboratory equipment and could produce very good chemical analyses, which are certified by a high number of international geochemical standards.

The significant part of geochemical analyses recently represents chemical analyses of rock-forming and especially accessory minerals. For these purposes are used scanning electron microscopes equipped by energy-dispersive X-ray detectors (EDS) or more better electron microprobe analysers (EMPA), which are equipped with EDS detectors and wavelength-dispersive X-ray detectors (WDS), which are more suitable for quantitative analyses of rare elements, inclusive of REE. Recently are predominantly used microprobes produced by CAMECA in France and JEOL in Japan. The main interest is usually concentrated on detailed chemical analyses of different accessory minerals, predominantly on analyses of zircon, monazite and xenotime. By study of granites coupled with different ore mineralisation (Sn-W, U) the microprobe analyses are concentrated also on chemistry of Sn-W-Nb-Ta minerals and different uranium- and thorium-bearing minerals.

The important part of geochemistry represents interpretation and presentation of geochemical data. For these purposes are recently used predominantly two different software packages, the commercial software Minpet [6] and free available software GCDKit [7]. The software Minpet could be used only with Windows, version 7, whereas software GCDKit could be used under Windows versions 7-11. This software offers distinctly more possibilities for presentation and interpretation chemical analyses of granites, together with their statistical analyses, determination of granite melt temperatures based on analyses of Zr and REE and basic principles of geochemical modelling igneous processes.

The predominantly used geochemical classifications of granites are based on the determination of A/CNK ratio (mol. $Al_2O_3/(CaO + Na_2O + K_2O)$. According this ratio are granites classified as peraluminous or metaluminous granites and/or as S- and I-type granites [8, 9]. Some other classifications of granites on I- and S-types are based on distribution of magnetite or ilmenite [10]. Similar classification of granitic rocks is based on modified alkali-lime index ($Na_2O + K_2O - CaO$) and content of SiO_2. According to this classification, the granites could be distinguished on alkalic, alkali-calcic, calc-alkalic and calcic granitic rock [11]. Some other geochemical family of granites is A-type granites (anorogenic) which are in detail classified using the determinations of Al, Ga, Nb, Ta, Y and Zr [12, 13].

4. Metallogeny

Granites are very often associated with different types of ore mineralisation, predominantly with U and Sn-W-Li bearing mineralisation. The mineralised granites are in these cases altered in different rock types, as well as aceites, episyenites and greisens. The basic internationally accepted terminology of these hydrothermally altered granitic rocks was adopted by the International Union of Geological Sciences (IUGS) [14].

The origin of aceites is coupled with uranium mineralisation originated in shear zones, which occurs in granitic rocks or in highly metamorphosed rocks (paragneisses). These rock series are usually altered in mixture of albite, chlorite and clay minerals, with different distributions of uranium minerals (uraninite, coffinite) (**Figure 4**). This low-temperature hydrothermal alteration is coupled with significant

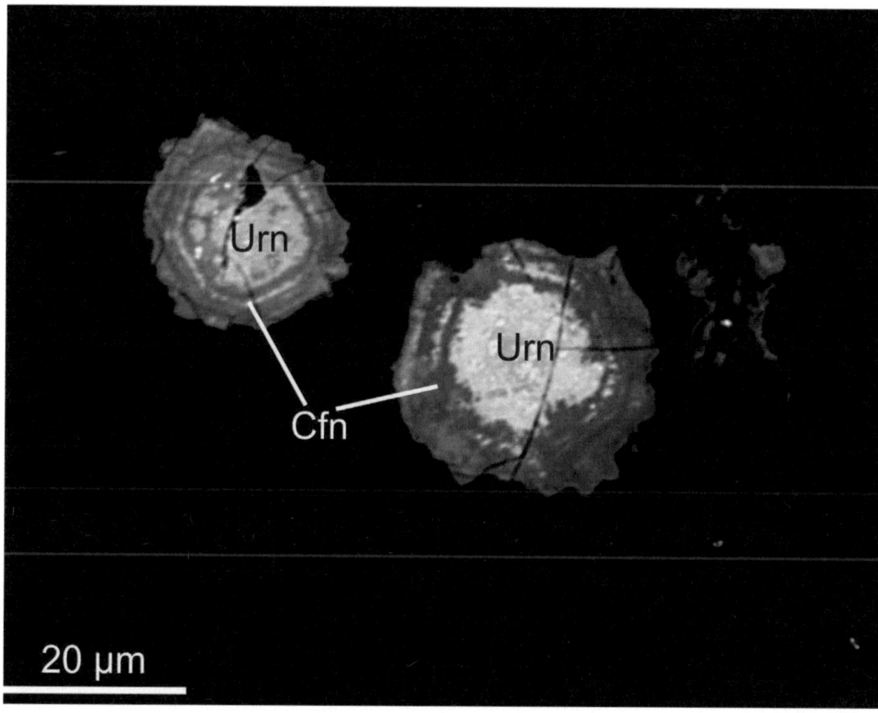

Figure 4.
Back scattered electron (BSE) image of uraninite and coffinite from uranium deposit Okrouhlá Radouň, Bohemian massif (urn—Uraninite, Cfn—Coffinite).

removal of original magmatic quartz. For describing of these altered rocks, evolved in disseminated uranium deposits in the Massif Central and Armorican Massif, France, were in the past used term episyenites [15, 16]. However, according to the recent IUGS classification for metasomatic rocks [14], the term episyenite could be abandoned. The term aceite was introduced to geosciences by Omel'yanenko [17].

The greisenisation is one of most significant hydrothermal alterations, which occurs in granite-related Sn-W ore deposits. Historically, the term 'greisen' has been used firstly by miners from the Krušné Hory/Erzgebirge Mts. to describe wall rocks consisting of quartz, Li-mica and topaz surrounding the Sn-W mineralisation, which occurs in this area [18]. This area hosts a number of Sn-W deposits (e.g. Cínovec/Zinnwald, Altenberg, Ehrenfriedersdorf, Krásno-Horní Slavkov) bound to greisenised granite stocks of the Variscan granite bodies. These granites represent highly fractionated, high-F, Li-mica granites of the Krušné Hory/Erzgebirge batholith [19].

5. Conclusions

The chapter discussed the composition of granites, together with used mineralogical and geochemical methods, which are recently used for description and discussion of their composition and origin. The mineralogical and geochemical methods, together with study of relation of granites to uranium and tin-tungsten ore deposits, are recently the most significant research methods used in geoscience.

Acknowledgements

The author would like to thanks to the support of the long-term conceptual development research organisation RVO: 67985891.

Conflict of interest

There is no conflict of interest.

Author details

Miloš René
Institute of Rock Structure and Mechanics, v.v.i., Czech Academy of Sciences, Prague, Czech Republic

*Address all correspondence to: rene@irsm.cas.cz

References

[1] Le Maitre RW, editor. Igneous Rocks. A Classification and Glossary of Terms. 2nd ed. Cambridge: Cambridge University Press; 2002. p. 236

[2] Müller A, Kerkhof van der AM, Behr HJ, Kronz A, Koch-Müller M. The evolution of late-Hercynian granites and rhyolites documented by quartz – A review. Earth and Environmental Science Transactions of the Royal Society of Edinburg. 2009;**100**:185-204. DOI: 10.1017/S1755691009016144

[3] Slaby E, Götze J, Wörner G, Simon K, Wrzalik R, Smigielski M. K-feldspar phenocrysts in microgranular magmatic enclaves: A cathodoluminescence and geochemical study of crystal growth as a marker of magma mingling dynamics. Lithos. 2008;**105**:85-97. DOI: 10.1016/j.lithos.2008.02.006

[4] Kempe U, Götze J. Cathodoluminescence (CL) behaviour and crystal chemistry of apatite from rare-metal deposits. Mineralogical Magazine. 2002;**66**:151-172. DOI: 10.1180/0026461026610019

[5] Rollinson HR. Using geochemical data: Evaluation, presentation, interpretation. Essex: Longham; 1993. p. 352

[6] Richard LR. MinPet: Mineralogica and Petrological Data Processing System, Version 2.02. Québec, Canada: MinPet Geological Software; 1988-1995

[7] Janoušek V, Moyen JF, Hervé M, Erban V, Farrow C. Geochemical Modelling of Igneous Processes – Principles and Recipes in R Language. Heidelberg: Springer. p. 346. DOI: 10.1007/978-3-662-46791-6

[8] Shand SJ. Eruptive Rocks. London: Thomas Murby& Co.; 1927. p. 360

[9] Chappell BW, White AJR. Two contrasting granite types. Pacific Geology. 1974;**8**:173-174

[10] Ishihara S. The magnetite-series and ilmenite-series granitic rocks. Mining Geology. 1977;**27**:293-305

[11] Frost BR, Frost CD. A geochemical classification for feldspathic igneous rocks. Journal of Petrology. 2008;**49**:1955-1969. DOI: 10.1093/petrology/cgn054

[12] Whalen JB, Currie KI, Chappell BW. A-type granites: Geochemical characteristics, discrimination and petrogenesis. Contribution to Mineralogy and Petrology. 1987;**95**:407-419. DOI: 10.1007/BF00462202

[13] Eby GN. Chemical subdivision of the A-type granitoids: Petrogenetic and tectonic implications. Geology. 1992;**20**:641-644. DOI: 10.1130/0091-7613(1992)020˂0637

[14] Fettes D, Desmons J, editors. Metamorphic Rocks. A Classification and Glossary of Terms. Cambridge: Cambridge University Press; 2007. p. 244

[15] Cathelineau M. The hydrothermal alkali metasomatism effects on granitic rocks. Quartz dissolution and related subsolidus changes. Journal of Petrology. 1986;**27**:945-965. DOI: 10.1093/petrology 127.4.945

[16] Leroy J. The Margnac and Fanay uranium deposits of the Le Crouzille district (western massif central, France): Geologic and fluid inclusion studies. Economic Geology. 1978;**73**:1611-1634. DOI: 10.2113/gsecongeo.73.8.1811

[17] Omelýanenko BI. Wall-Rock Hydrothermal Alterations. Moscow: Nedra Publishing House; 1978. p. 215 (in Russian)

[18] Lempe JF. Beschreibung des Bergbaues aus dem Sächsiscen Zinnwalde. Magazin für die Bergbaukunde. 1785;**1**:100-142

[19] Förster HJ, Tischendorf G, Trumbull RB, Gottesmann B. Late-collisional granites in the Variscan Erzgebirge, Germany. Journal of Petrology. 1999;**40**:1613-1645. DOI: 10.1093/petroj/40.11.1613

Chapter 2

Granitoids of the Mauthausen Type in the Czech Part of the Moldanubian Batholith

Miloš René and Zdeněk Dolníček

Abstract

The Moldanubian Batholith is the largest Variscan magmatic complex in the Bohemian Massif, which is part of the Central European Hercynian belt. In northern part of the Moldanubian Batholith occur relatively small bodies of granitoids which could be correlated with biotite granodiorites of the Mauthausen type which occur in the Austrian part of this batholithic complex. The first body is formed by biotite-muscovite granite of the Pavlov type. The second occurrence of granitoids of the Mauthausen type is formed by two, relatively small bodies of the biotite granodiorites of the Pohled type. The estimation of melting temperatures of granitic melts for granitic rocks from Pavlov and Pohled area, based on zircon and monazite saturation thermometers show that melting temperatures were partly higher than those of the Mauthausen granodiorites the Austrian part of the Moldanubian Batholith (732–817°C). Analysed apatites from both areas contain high F (3.05–4.00 wt.%) and little Cl (0.00–0.06 wt.%). The analysed zircons contain low Hf concentrations (0.93–1.65 wt.% HfO_2, 0.008–0.013 apfu Hf). The analysed monazites form the Pavlov and Pohled granitoids plot close to the huttonite vector.

Keywords: granite, Moldanubian batholith, bohemian massif, petrology, geochemistry

1. Introduction

The biotite granodiorites of the Mauthausen type represent the youngest group of granitic rocks occurred in the Moldanubian Batholith, that are part of the Freistadt/Mauthausen suite (**Figure 1**) [1–3]. The granodiorites of the Mauthausen type occur predominantly in the Austrian part of the Moldanubian Batholith, especially along of the Danube River near of the Mauthausen town. Petrology and geochemistry of these granodiorites were in detail described by Vellmer and Wedepohl [4] and by Gerdes [5]. Smaller occurrences of this granodiorites exist also in the Austrian Mühlviertel. Later were found occurrences of granodiorites of the Mauthausen type also in northern part of the Moldanubian Batholith near Humpolec and Havlíčkův Brod in the

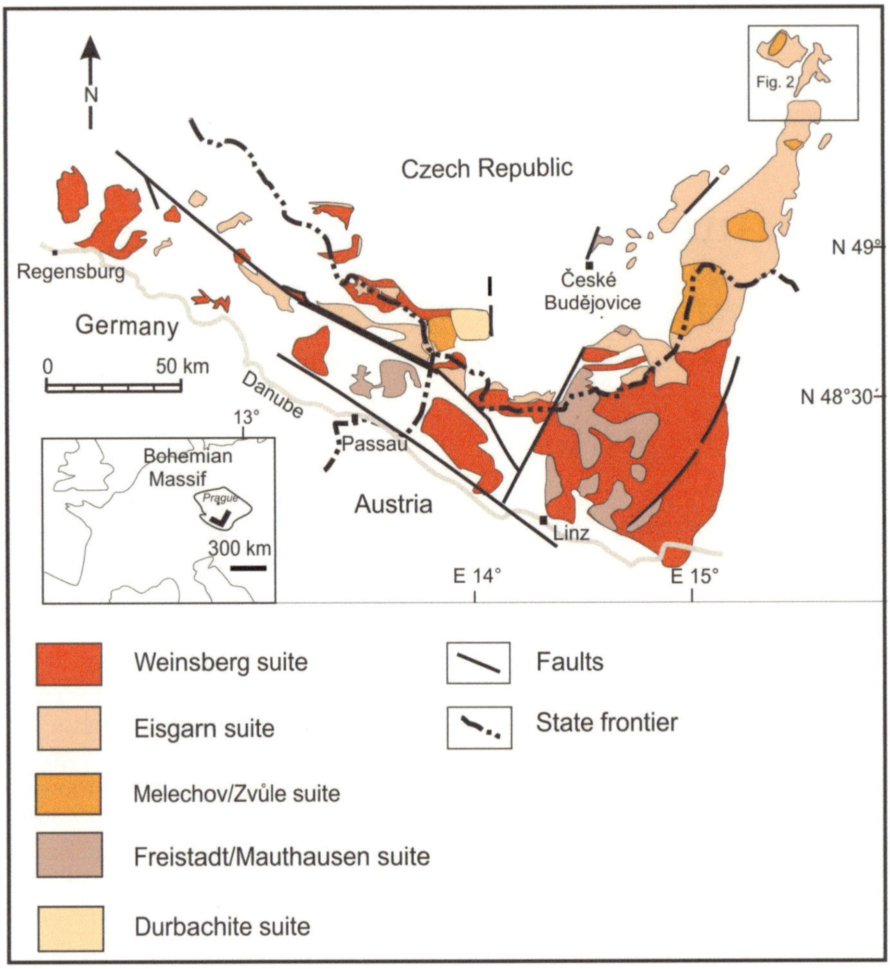

Figure 1.
Geological map of the Moldanubian Batholith (after [1, 2], modified by authors).

Czech Republic (**Figure 2**) [6–9]. However, in papers of Janoušek, Matějka and René, Matějka [7, 8] were these occurrences and their compositions described only very briefly. Therefore, the aim of presented chapter is detailed petrology, mineralogy and geochemistry of these granitoids, together with discussion of actually presented position of the Moldanubian Batholith in the Central European Variscan belt as whole [3, 10]. The similarity of granitoids of the Mauthausen type occurring in northern part of the Moldanubian Batholith with the Mauthausen granodiorites occurred in Austria is based on detailed geochemical study these granitoids and composition of selected accessory minerals, predominantly monazite.

2. Geological setting

The Moldanubian Batholith (and/or also South Bohemian Batholith) together with Fichtelgebirge/Erzgebirge Batholith on the basis of the synchronicity of geochronological data and the similarity of granite types belongs to one coherent and cogenetic

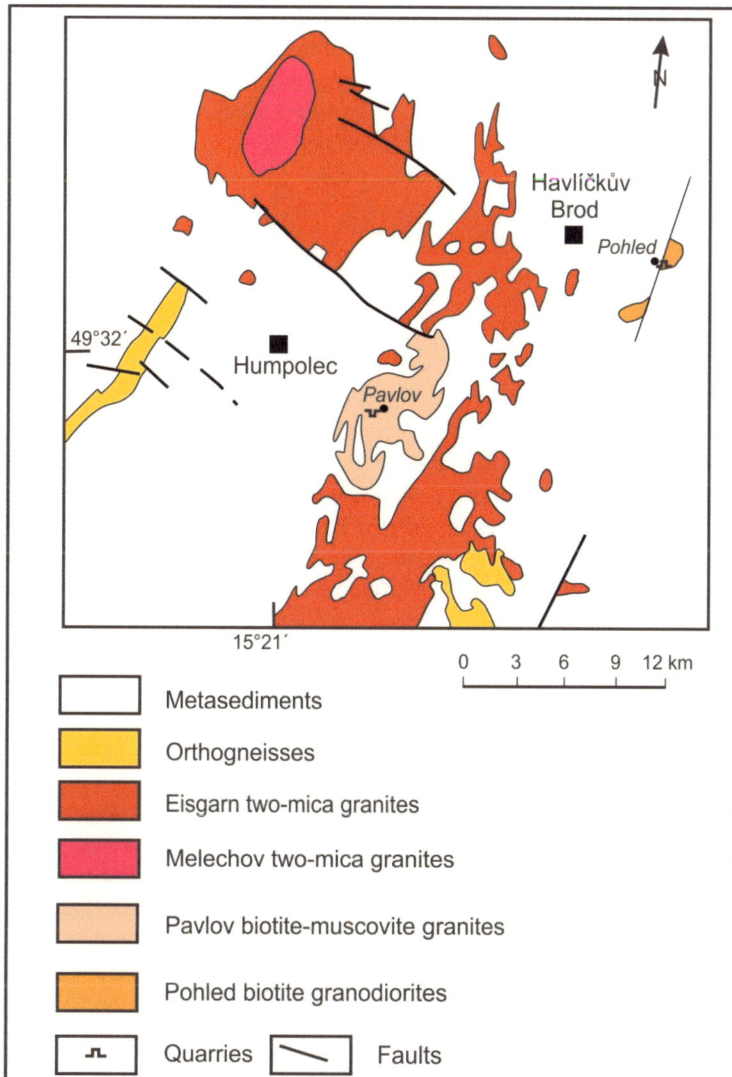

Figure 2.
Geological map of the northern part of the Moldanubian Batholith (after [6], modified by authors).

plutonic megastructure, the Saxo-Danubian granite belt evolved in the Central European Variscides [10]. The granitoids of the Mauthausen type, together with the Freistadt granodiorites are part of the youngest magmatic group (318–316 Ma) of the Moldanubian Batholith [3, 10]. This magmatic group is recently interpreted as result of renewed decompression melting [3]. Dating of the Mauthausen granodiorites (316 ± 1 Ma, monazite, U-Pb dating by isotope dilution thermal ion mass spectrometry) is based on dating one sample from the Mauthausen quarry in Austria [10].

The granitoids of the Mauthausen type occurring in northern part of the Moldanubian Batholith are represented by three relatively small bodies of the biotite-muscovite granites of the Pavlov type and biotite granodiorites of the Pohled type (**Figure 2**). The fine-grained biotite-muscovite granites of the Pavlov type were firstly in detail mapped and recognised as individual granite variety in the northern part of the Moldanubian Batholith by Veselá et al. [11]. Its petrology and geochemistry were later described in detail by Matějka and Janoušek [7]. The biotite-muscovite granite of the Pavlov type occurs as NNE-SSW elongated body which was in the past opened by small quarries near Pavlov and Slavníč villages. During geological mapping two small biotite granodiorite bodies were recognised in quarries by villages Vysoká and Pohled, near Havlíčkův Brod [12]. The petrology and geochemistry of biotite granodiorite from the quarry Pohled was later briefly described by Mastíková [9] and Doleželová [13].

3. Materials and methods

Representative rock samples weighting 2–5 kg, collected from quarries Pavlov and Slavníč were crushed in a jaw crusher and representative split of these samples were ground in an agate ball mill. Major and trace elements were determined by inductively coupled plasma mass spectrometry (ICP-MS) techniques at Activation Laboratories Ltd., Ancaster, Canada, using a Perkin Elmer Sciex ELAN 6100 ICP mass spectrometer, following standard lithium metaborate/tetraborate fusion and acid decomposition of the sample. Similar analytical techniques were used also for chemical analyses of four representative rock samples of biotite granodiorite from the Pohled quarry, which were performed in ACME laboratory in Vancouver, Canada. All analyses were calibrated against international reference materials. Geochemical data for the Mauthausen biotite granodiorite were taken from paper of Gerdes [5].

Approximately 140 quantitative electron microprobe analyses of apatite, zircon, monazite and selected rock-forming minerals (plagioclase, K-feldspar and biotite) were collected from representative samples of the Pavlov, Pohled and Mauthausen granitoids. All these minerals were analysed in polished thin sections. The back-scattered electron (BSE) images were acquired to study the internal structure of mineral aggregates and individual mineral grains. The abundances of all chemical elements were determined using a CAMECA SX 100 electron probe micro-analyser (EPMA) operated in wavelength-dispersive mode at the Department of Geological Sciences, Masaryk University in Brno and in National Museum, Prague. The accelerating voltage and beam currents were 15 kV and 20 nA or 40 nA, respectively, and the beam diameter was 1 to 5 μm. The peak count time was 20 s, and the background time was 10 s for major elements. For the minor elements, the counting times were 40–60 s on the peaks, and 20–30 s on each background position. The following standards, X-ray lines and crystals (in parentheses) were used: AlK_α – sanidine (TAP), CaK_α – fluorapatite (PET), CeL_α – $CePO_4$ (PET), ClK_α – vanadinite (LPET), DyL_α – $DyPO_4$ (LLIF), ErL_α – $ErPO_4$ (PET), EuL_β – (LLIF), FK_α – topaz (PC1), FeK_α – almandine (LLIF), GdL_β – $GdPO_4$ (LLIF), HfM_α – Hf (TAP), KK_α – sanidine (TAP), LaL_α – $LaPO_4$ (PET), MgK_α –Mg_2SiO_4 (TAP), MnK_α –spessartine (LLIF), NaK_α – albite (PET), NbL_α – columbite, Ivigtut (LPET), NdL_β – $NdPO_4$ (LLIF), PK_α – fluorapatite (PET), PbM_α – vanadinite (PET), PrL_β – $PrPO_4$ (LLIF), RbL_α –RbCl (LTAP), SK_α – $SrSO_4$ (LPET), ScKα – ScP_5O_{14} (PET), SiK_α – sanidine (TAP), SmL_β – $SmPO_4$ (LLIF), SrL_α – $SrSO_4$ (TAP), TaM_α – $CrTa_2O_6$ (TAP), TbL_α – $TbPO_4$ (LLIF), ThM_α – $CaTh(PO_4)_2$ (PET), TiK_α – anatase (PET), UM_β –metallic U (PET), VK_β – vanadinite (LPET), YL_α – YPO_4 (PET), YbL_α –$YbPO_4$ (LLIF) and ZrL_α – zircon (TAP). The raw data were corrected using the PAP matrix corrections [14]. The detection limits were approximately 400–500 ppm for Y, 600 ppm for Zr, 500–800 ppm for REE and 600–

700 ppm for U and Th. Mineral formulae were recalculated using the MinPet 2.02 software [15]. The calculation of mineral formulae for end-member F-, Cl- and OH-apatites was performed according to Piccoli and Candela [16]. Mole fractions for components in monazite and xenotime were calculated according to Pyle et al. [17].

4. Results

4.1 Petrology

Biotite-muscovite granites of the Pavlov type are fine grained, usually equigranular rocks. Major components of this granite are quartz (29–35 vol.%), plagioclase (An_{22-37}) (25–31 vol.%), K-feldspar (21–30 vol.%), biotite (5–9 vol.%) and muscovite (2–4 vol.%) (**Figure 3**). Biotite is represented by annite (Fe/Fe + Mg = 0.60–0.63, Al^{4+} = 2.22–2.23 apfu and Ti = 0.32–0.36 apfu (atoms per formula unit)). Accessory minerals are represented by apatite, zircon, ilmenite and monazite.

Biotite granodiorites of the Pohled type are fine to medium grained, usually equigranular rocks. Major components of this granodiorite are plagioclase (An_{16-45}) (40–48 vol.%), quartz (24–37 vol.%), K-feldspar (12–15 vol.%) and biotite (10–13 vol.%) (**Figure 4**). Biotite is represented by annite (Fe/Fe + Mg = 0.51–0.54, Al^{4+} = 2.05–2.21 apfu and Ti = 0.45–0.55 apfu (atoms per formula unit)). Accessory minerals are represented by apatite, zircon, ilmenite and monazite. In places also hydrothermal allanite was found.

Figure 3.
Microphotograph of the Pavlov biotite-muscovite granite (Bt – Biotite, Kfs – K-feldspar, Ms. – Muscovite, Pl – Plagioclase, Qz – Quartz), thin section, crossed polarizers.

Figure 4.
Microphotograph of the Pohled biotite granodiorite (Bt – Biotite, Kfs – K-feldspar, Pl – Plagioclase, Qz – Quartz), thin section, crossed polarizers.

4.2 Geochemistry

Biotite-muscovite granites of the Pavlov type are weakly peraluminous granites with A/CNK [mol. $Al_2O_3/(CaO + Na_2O + K_2O)$] of 1.15–1.22 and contents of SiO_2 68.5–69.9 wt.%, Na_2O 3.3–3.8 wt.% and K_2O 3.9–4.1 wt.%. These granites display relatively poorly fractionated REE pattern (La_N/Yb_N = 16.2–21.1) and negative Eu anomaly (Eu/Eu^* = 0.33–0.54). In comparison with more fractionated two-mica granites of the Eisgarn type, the granites of the Pavlov type are enriched in CaO (1.9–2.1 wt.%), Ba (791–811 ppm), Sr. (505–535 ppm) and depleted in Rb (194–207 ppm).

Biotite granodiorites of the Pohled type are weakly peraluminous granites with A/CNK 1.05–1.15 and contents of SiO_2 66.9–68.0 wt.%, Na_2O 3.2–3.5 wt.% and K_2O 3.7–4.3 wt.%, relatively poorly fractionated REE pattern (La_N/Yb_N = 12.3–23.5) and gently negative to missing Eu anomaly (Eu/Eu^* = 0.73–1.05). In comparison with more fractionated two-mica granites of the Eisgarn type, the granodiorites of the Pohled type are distinctly enriched in CaO (2.0–2.9 wt.%), Ba (829–951 ppm), Sr. (603–668 ppm) and depleted in Rb (147–172 ppm) (**Table 1**).

4.3 Accessory minerals association

The REE, Y and Zr bearing accessories in granites and granodiorites of the Pavlov and Pohled types are represented by primary magmatic apatite, zircon and monazite. In biotite granodiorites of the Pohled type also hydrothermal allanite was found. Apatite occurs usually in form of euhedral and subhedral grains (20–50 μm in size).

Sample	**R-1305**	**R-1306**	**Do-1**	**Do-3**
Locality	**Slavníč**	**Pavlov**	**Pohled**	**Pohled**
Rock **wt %**	**Biotite-muscovite granite**	**Biotite-muscovite granite**	**Biotite granodiorite**	**Biotite granodiorite**
SiO_2	68.51	69.85	67.48	68.02
TiO_2	0.40	0.38	0.50	0.48
Al_2O_3	16.64	16.14	15.70	15.62
Fe_2O_3 tot.	2.29	2.14	3.40	3.21
MnO	0.03	0.03	0.06	0.05
MgO	0.65	0.68	1.30	1.22
CaO	2.12	1.92	2.59	2.32
Na_2O	3.81	3.32	3.28	3.34
K_2O	4.06	3.92	3.95	3.83
P_2O_5	0.18	0.16	0.15	0.15
CO_2	0.04	0.04	0.04	0.08
S	0.01	0.01	0.05	0.04
L.O.I.	0.80	0.96	1.30	1.50
Total	99.54	99.55	99.80	99.86
A/CNK	1.15	1.22	1.09	1.13
ppm				
Ba	811	791	951	829
Rb	194	207	158	147
Sr	535	505	628	603
Zr	160	162	139	141
U	3.5	3.3	8.5	5.5
Th	18.4	18.0	14.9	12.2

Table 1.
Representative compositions of the Pavlov and Pohled granitoids.

Zircon usually occurs as small euhedral to subhedral grains (10–80 μm in size). Some zircon grains are oscillatory zoned (**Figure 5**). Monazite occurs as relatively rare, usually subhedral to anhedral grains (20–30 μm). The compositions of apatite, zircon and monazite were studied in detail.

4.3.1 Apatite composition

All analysed apatites contain high F (3.05–4.00 wt.%) and low Cl (0.00–0.06 wt.%). Their contents of Fe (0.14–0.49 wt.% FeO) and Mn (0.11–0.35 wt.% MnO) are low. Their contents of sulphur and natrium are low (0.00–0.05 wt.% SO_3, 0.00–0.09 wt.% Na_2O). The concentrations of Y are low (0.00–0.14 wt.% Y_2O_3) (**Table 2**).

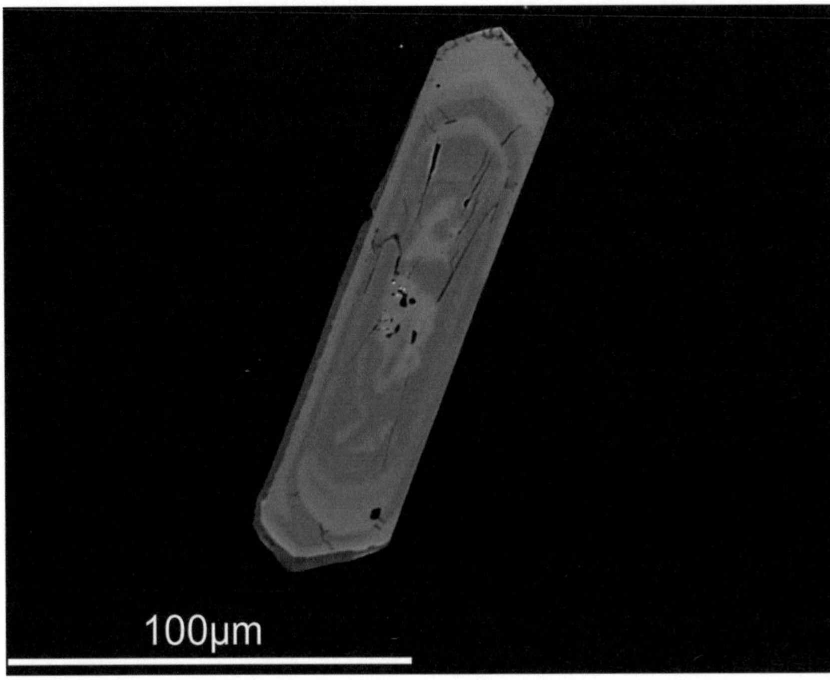

Figure 5.
BSE picture of oscillatory zoned zircon from the Pohled biotite granodiorite.

Sample	**1305–24**	**1305–25**	**1306–12**	**1306–13**	**Po-G1–97**	**Po-G2–91**
Locality	**Slavníč**	**Slavníč**	**Pavlov**	**Pavlov**	**Pohled**	**Pohled**
P_2O_5	41.51	41.41	42.03	42.09	41.33	41.74
SiO_2	0.00	0.14	0.12	0.11	0.00	0.16
Y_2O_3	b.d.l.	0.04	0.11	b.d.l.	b.d.l.	b.d.l.
CaO	55.04	54.31	54.76	55.06	55.04	55.08
FeO	0.20	0.38	0.32	0.47	0.10	0.06
MnO	0.11	0.27	0.28	0.12	0.14	0.09
Na_2O	0.00	0.00	0.00	0.44	0.00	0.00
SO_3	0.00	0.05	0.00	0.00	0.06	0.05
F	2.85	3.66	3.37	3.75	3.77	3.92
Cl	0.02	0.01	0.00	0.00	0.09	0.17
O=F,Cl	−1.20	−1.54	−1.42	−1.58	−1.61	−1.69
Total	98.53	98.73	99.57	100.46	98.92	99.58
X_{Fap}	0.757	0.972	0.895	0.995	0.987	0.977
X_{Clap}	0.003	0.001	0.000	0.000	0.013	0.023
X_{OHap}	0.240	0.027	0.105	0.005	0.000	0.000

b.d.l. below detection limit.

Table 2.
Representative microprobe analyses of apatite (wt. %).

4.3.2 Zircon composition

The analysed zircons contain low Hf concentrations (0.93–1.65 wt.% HfO_2, 0.008–0.013 apfu Hf, **Table 3, Figure 6**). The atomic ratio Hf/(Hf + Zr) varies from 0.008 to 0.014. The concentrations of Y in analysed zircons are relatively low (up to 1.08 wt.% Y_2O_3, 0.018 apfu Y). All analysed zircons display also low concentrations of U and Th, reached up to 0.52 wt.% UO_2, 0.004 apfu U and up to 0.88 wt.% ThO_2, 0.006 apfu Th in zircons from the Pohled biotite granodiorite. The concentrations of both elements

Sample	**1305–13**	**1305–16**	**1306–6**	**1306–7**	**Po-G124**	**Po-G1–25**
Locality	**Slavníč**	**Slavníč**	**Pavlov**	**Pavlov**	**Pohled**	**Pohled**
SiO_2	32.30	32.78	31.52	31.92	32.69	32.55
Al_2O_3	0.00	0.10	0.01	0.00	0.00	0.00
ZrO_2	64.69	65.20	64.89	65.78	65.85	64.72
HfO_2	1.06	1.04	1.04	1.00	1.25	1.17
CaO	0.00	0.08	0.01	0.01	0.03	0.02
FeO	0.44	0.01	0.53	0.72	0.01	0.01
P_2O_5	0.00	0.00	0.40	0.25	0.00	0.00
Sc_2O_3	0.08	0.05	0.10	0.05	0.00	0.06
Y_2O_3	0.26	0.03	0.35	0.00	0.00	0.07
Yb_2O_3	0.16	0.09	0.20	0.12	0.02	0.12
UO_2	0.12	0.16	b.d.l.	b.d.l.	b.d.l.	b.d.l.
ThO_2	0.13	0.08	0.15	0.11	b.d.l.	0.51
Total	99.24	99.62	99.20	99.96	99.85	99.23
apfu, O = 4						
Si	0.999	1.005	0.976	0.981	1.002	1.004
Al	0.000	0.004	0.000	0.000	0.000	0.000
Zr	0.976	0.975	0.980	0.985	0.984	0.973
Hf	0.009	0.009	0.009	0.009	0.011	0.010
Ca	0.000	0.003	0.000	0.000	0.001	0.001
Fe	0.011	0.000	0.014	0.019	0.000	0.000
P	0.000	0.000	0.010	0.007	0.000	0.000
Sc	0.002	0.001	0.003	0.001	0.000	0.002
Y	0.004	0.000	0.006	0.000	0.000	0.001
Yb	0.002	0.001	0.002	0.004	0.000	0.001
U	0.001	0.001	0.000	0.000	0.000	0.003
Th	0.001	0.001	0.001	0.001	0.000	0.004

b.d.l. – below detection limit.

Table 3.
Representative microprobe analyses of zircon (wt. %).

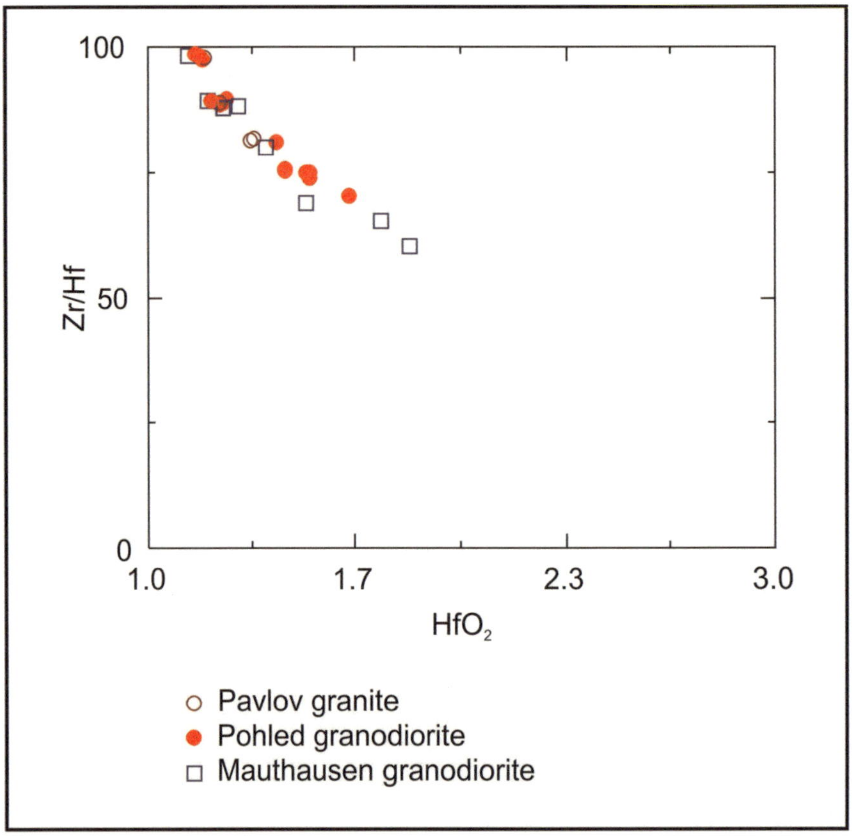

Figure 6.
Chemical composition of zircon.

in biotite-muscovite granites of the Pavlov type are partly lower (up to 0.28 wt.% UO_2, 0.002 apfu U; up to 0.18 wt.% ThO_2, 0.001 apfu Th).

4.3.3 Monazite composition

In the Pavlov and Pohled granitoids, the sum of LREE (La + Ce + Pr + Nd + Sm) ranges between 3.27 and 3.87 apfu (calculation is based on 16 atoms of oxygen), being relatively higher in the Pohled biotite granodiorites. Cerium is in all cases the most abundant REE varying between 27.16 and 33.71 wt.% Ce_2O_3 (1.63–1.97 apfu Ce). The second most abundant REE is La 10.64–20.00 wt.% La_2O_3, 0.62–1.18 apfu La), followed by Nd (8.85–12.38 wt.% Nd_2O_3, 0.50–0.70 apfu Nd), Pr (2.54–3.42 wt.% Pr_2O_3, 0.15–0.20 apfu Pr) and Sm (0.76–2.41 wt.% Sm_2O_3, 0.04–0.13 apfu Sm). Thus, all analysed monazite grains form the Pavlov and Pohled granitoids should be termed monazite-(Ce) (**Table 4**). However, the ranges in atomic ratios amongst individual REEs vary broadly, the $(La/Nd)_{cn}$ ratio is between 1.65 and 4.35, the $(La/Sm)_{cn}$ ratio is between 3.06 and 14.57. The content of Y in analysed monazites from the Pavlov and Pohled granitoids ranges between 0.23 and 2.67 wt.% Y_2O_3 (0.02–0.22 apfu Y). The concentrations of Th vary between 0.14 and 9.54 wt.% ThO_2 (0.01–0.35 apfu Th). The concentrations of U vary between below detection limit and 0.99 wt.% UO_2 (0.00–0.04 apfu U). Two main coupled substitution mechanism have been proposed for monazite,

Sample	**1305–11**	**1305–14**	**1306–22**	**1306–25**	**Po-G1–22**	**Po-G1–27**
Locality	**Slavníč**	**Slavníč**	**Pavlov**	**Pavlov**	**Pohled**	**Pohled**
P_2O_5	28.54	28.06	27.91	29.25	29.68	28.28
SiO_2	1.19	1.08	1.32	0.78	0.44	0.94
ThO_2	7.18	6.44	5.74	4.30	2.98	3.67
UO_2	0.22	0.16	0.15	0.09	0.09	0.00
Y_2O_3	0.46	0.50	0.23	0.42	2.67	0.59
La_2O_3	13.89	14.53	16.84	16.10	13.59	16.88
Ce_2O_3	30.02	30.60	31.27	31.84	28.85	30.90
Pr_2O_3	3.12	2.99	2.93	3.16	3.25	3.00
Nd_2O_3	10.89	11.04	10.37	11.06	11.77	9.69
Sm_2O_3	1.48	1.41	0.99	1.45	2.13	1.37
Gd_2O_3	0.60	0.68	0.41	0.80	1.29	0.56
Dy_2O_3	0.20	0.14	b.d.l.	0.12	0.70	0.27
Er_2O_3	b.d.l.	b.d.l.	b.d.l.	b.d.l.	0.28	0.16
Yb_2O_3	b.d.l.	b.d.l.	b.d.l.	b.d.l.	0.12	0.08
CaO	0.78	0.63	0.75	0.54	0.59	0.69
PbO	0.09	0.09	0.10	0.06	0.03	0.10
Total	98.66	98.35	99.01	99.97	98.46	97.18
apfu, O = 16						
P	3.849	3.821	3.780	3.892	3.953	3.864
Si	0.190	0.174	0.211	0.123	0.069	0.152
Th	0.260	0.236	0.209	0.154	0.107	0.135
U	0.008	0.006	0.005	0.003	0.003	0.000
Y	0.039	0.043	0.020	0.035	0.223	0.051
La	0.815	0.861	0.993	0.932	0.788	1.004
Ce	1.749	1.800	1.829	1.830	1.660	1.824
Pr	0.181	0.175	0.171	0.181	0.186	0.176
Nd	0.619	0.634	0.592	0.620	0.661	0.558
Sm	0.081	0.078	0.055	0.078	0.115	0.076
Gd	0.032	0.036	0.022	0.042	0.067	0.030
Dy	0.010	0.007	0.000	0.006	0.035	0.014
Er	0.000	0.000	0.000	0.000	0.014	0.008
Yb	0.000	0.000	0.000	0.000	0.006	0.004
Ca	0.133	0.109	0.129	0.091	0.099	0.119
Pb	0.004	0.004	0.004	0.003	0.001	0.004
Mole fractions:						
$LREEPO_4$	0.8759	0.8892	0.9014	0.9153	0.8585	0.9084
$HREEPO_4$	0.0112	0.0110	0.0077	0.0128	0.0325	0.0145

Sample	1305–11	1305–14	1306–22	1306–25	Po-G1–22	Po-G1–27
Locality	**Slavníč**	**Slavníč**	**Pavlov**	**Pavlov**	**Pohled**	**Pohled**
$CaTh(PO_4)_2$	0.0676	0.0546	0.0639	0.0458	0.0498	0.0594
$ThSiO_4$	0.0353	0.0343	0.0220	0.0173	0.0030	0.0050
YPO_4	0.0099	0.0108	0.0050	0.0088	0.0561	0.0127

b.d.l. – below detection limit.

Table 4.
Representative microprobe analyses of monazite (wt. %).

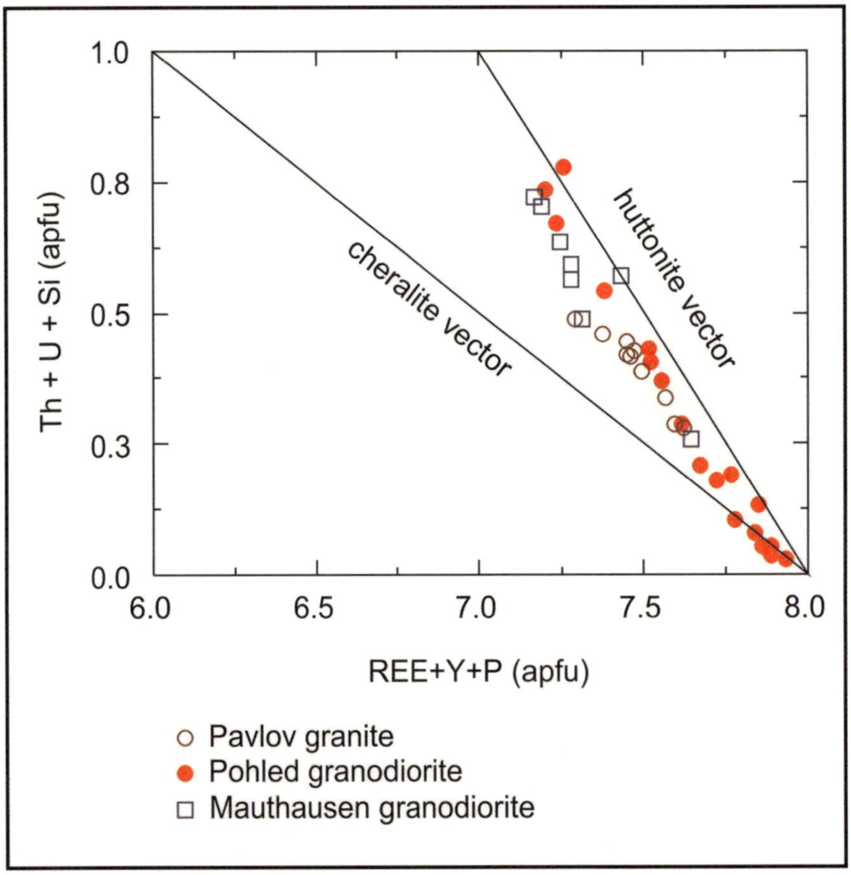

Figure 7.
Chemical composition of monazite.

namely the cheralite and huttonite substitutions [18]. The analysed monazites form the Pavlov and Pohled granitoids plot close to the huttonite vector (**Figure** 7).

5. Discussion

In the past, the origin and fractionation of granitic rocks of the Moldanubian Batholith was discussed by geochemical modelling based on trace-element

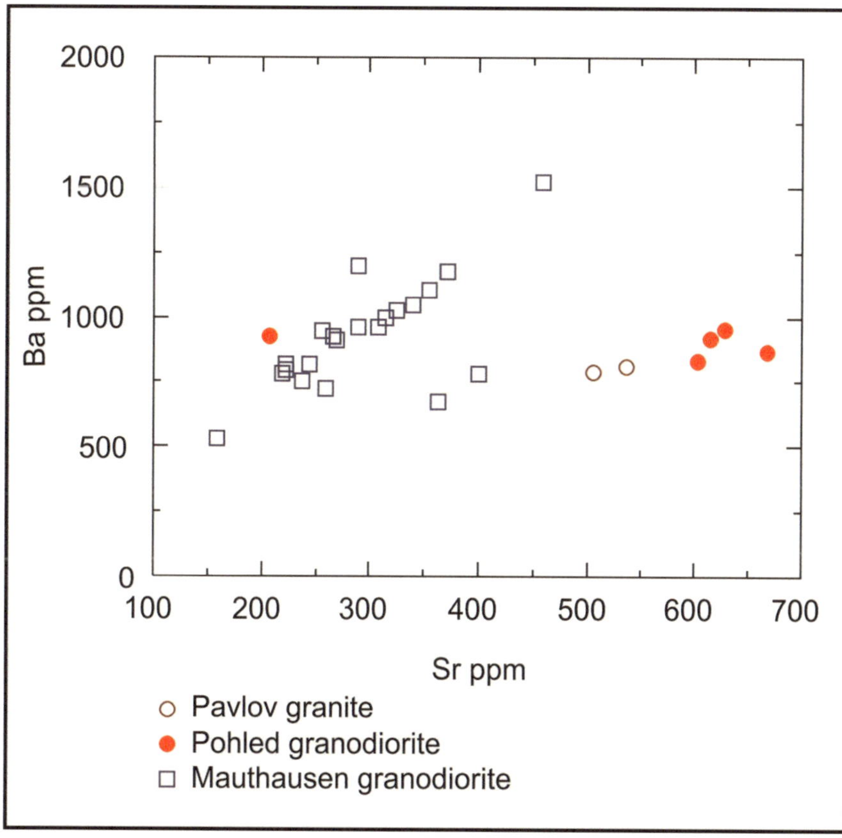

Figure 8.
Distribution of Ba an Sr in analysed rocks.

fractionation. According to these studies, granitic rocks of the Moldanubian Batholith could originate by LP-HT partial melting of various metasediments and/or by melting of a mixture of metasediments and amphibolites [7, 19–21]. According to majority of these studies, the granitic rocks of the individual magmatic suites occurred in the Moldanubian Batholith were also variably fractionated [7, 21]. The fractionation of these magmatic suites could be well documented by distribution of some trace elements (e.g., Ba, Sr., Th, Zr, REE) (**Figures 8** and **9**).

For distinguishing source rock series (greywackes vs. pelites) could be used some major elements, especially CaO/Na_2O and Al_2O_3/TiO_2 ratios [22]. According to these studies, in detail discussed by René [19], the granitic rocks of the Eisgarn suite originated by partial melting of metapelites, whereas granites and granodiorites of the Weinsberg and Freistadt/Mauthausen suites originated by partial melting of a metagreywackes-metabasalt mixture.

The estimation of melting temperatures of granitic melts is usually based on saturation thermometers based on melting of zircon and monazite [23–25]. According to zircon saturation thermometry granitic rocks are usually divided on the hot and cold [26]. The most detailed study of all problems connected with using of zircon thermometry was published by Siégel et al. [27] and Clemens et al. [28]. For all granitic rocks from the Moldanubian Batholith the T_{Zrnsat} was calculated according to revisited

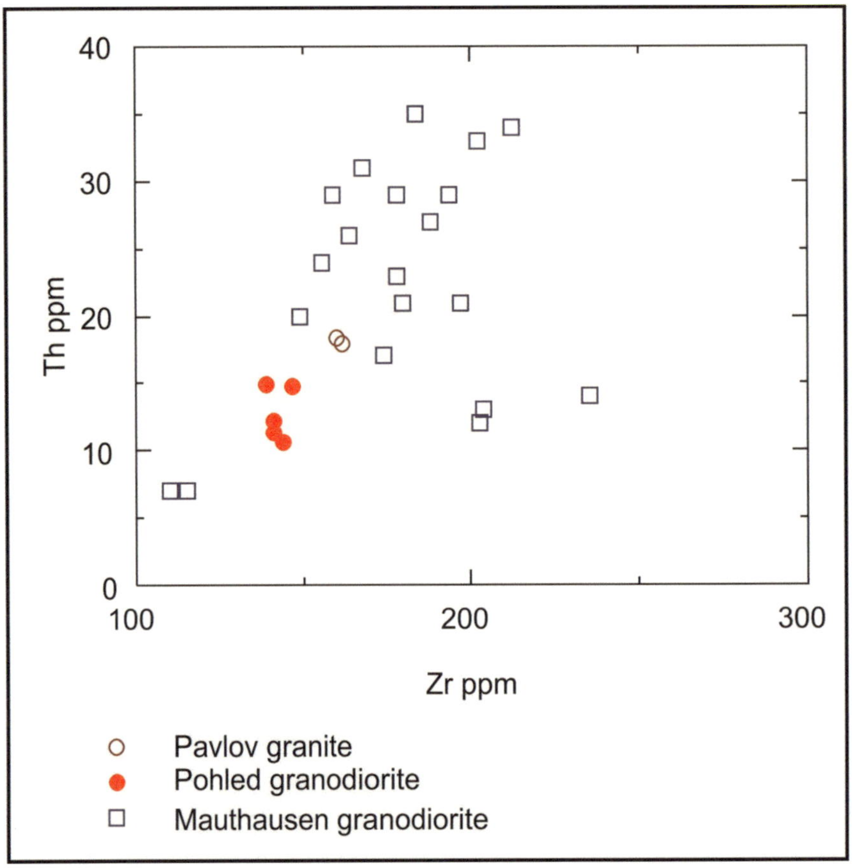

Figure 9.
Distribution of Th and Zr in analysed rocks.

formula published by Boehnke et al. [24] and T_{Mnzsat} according to model of Montel [25]. The saturation temperatures from both models for the Mauthausen granodiorite from the Austrian part of the Moldanubian Batholith varied between 693 and 803°C. However, in recent study of thermometry of the Moldanubian Batholith granitoids [3], based on older zircon saturation thermometry [23], the melting temperatures for the Mauthausen granites from occurrences in Austria are estimated in distinctly higher range (790–840°C). However, the new data of Ti-zircon thermometry for biotite granodiorite from the Mauthausen quarry give temperature data partly similar to our data presented in this study (736–844°C). The saturation temperatures for biotite-muscovite granites of the Pavlov type are partly higher (745–817°C). The saturation temperatures for the Pohled biotite granodiorite are similar (732–802°C). It is also interesting that in all these cases the T_{Mnzsat} is usually partly higher than the T_{Zrnsat} temperatures. These differences could be partly explained by restitic (inherited) monazite crystals from original metasediments.

The distribution of Fe, Mn, F and Cl in analysed apatites is similar to those of the S-type granites [15]. The low content of Hf in analysed zircon (0.93–1.65 wt.% HfO_2) is partly lower as its content in two-mica granites of the Eisgarn suite (1.0–2.5 wt.% HfO_2) [29]. The composition of analysed monazite is similar to those of the Freistadt biotite granodiorites, but distinctly different as its composition from two-mica granites of the Eisgarn suite. For monazites from the Freistadt granodiorites is similar huttonite substitution significant, whereas monazites from two-mica granites of the Eisgarn suite display cheralite substitution [30].

6. Conclusions

Granitic rocks of the Mauthausen type from northern part of the Moldanubian Batholith occurring between Pavlov and Pohled are according to their mineralogical and geochemical composition similar to the Mauthausen type occurrences in the Austrian part of the Moldanubian Batholith. These biotite-muscovite granites and biotite granodiorites are weakly peraluminous rocks, enriched especially in Ba and Sr. Their fractionation is documented by distribution of the Ba, Sr., Th and Zr. These granites and granodiorites originated by partial melting of a metagreywacke-metabasalt mixture. The estimation of melting temperatures of granitic melts for granitic rocks from Pavlov and Pohled area, based on zircon and monazite saturation thermometers, show that melting temperatures were partly higher as the melting temperatures for the Mauthausen granodiorites from the Austrian part of the Moldanubian Batholith (732–817°C). Analysed apatites from both areas contain high F (3.05–4.00 wt.%) and negligible Cl (0.0–0.06 wt.%). The analysed zircons contain low Hf concentrations (0.9–1.65 wt.% HfO_2, 0.00–0.013 apfu Hf). The composition of monazites form the Pavlov and Pohled granitoids plot close to the huttonite vector.

Acknowledgements

The support of the Long-Term Conceptual Development Research Organisation RVO 67985891 is thanked, Z.D. acknowledge financial support of the Ministry of Culture of the Czech Republic (long-term project DKRVO 2019–2023/1.I.e; National Museum, 00023272). We are grateful to R. Škoda and J. Haifler for their technical assistance by using electron microprobe analyses of selected minerals (plagioclase, biotite, zircon, monazite). We are also grateful for constructive comments of an anonymous reviewer and to F. Finger for new, unpublished data of Ti-zircon thermometry of the Mauthausen granite.

Conflict of interest

The authors declare no conflict of interests.

Author details

Miloš René[1*] and Zdeněk Dolníček[2]

1 Institute of Rock Structure and Mechanics, v.v.i., Czech Academy of Sciences, Prague, Czech Republic

2 National Museum, Prague, Czech Republic

*Address all correspondence to: rene@irsm.cas.cz

References

[1] Siebel W, Shang CK, Reiter E, Rohrmüller J, Breiter K. Two distinctive granite suites in the SW bohemian massif and their record of emplacement: Constraints from geochemistry and zircon $^{207}Pb/^{208}Pb$ chronology. Journal of Petrology. 2008;**49**:1853-1872. DOI: 10.1093/petrology/egn049

[2] Verner K, Žák J, Pertoldová J, Šrámek J, Trubač J, Týcová P. Magmatic history and geophysical signature of a post-collisional intrusive Centre emplaced near a crustal-scale shear zone: The Plechý granite pluton (Moldanubian batholith, bohemian massif). International Journal of Earth Science. 2009;**98**:517-532. DOI: 10.1007/s00531-007-0285-9

[3] Finger F, Schiller D, Lindner M, Hauzenberger C, Verner K, Žák J. Ultrahigh-temperature granites and curious thermal eye in the post-collisional south bohemian batholith of the Variscan orogenic belt (Europe). Geology. 2022;**50**:542-546. DOI: 10.1130/G49645.1

[4] Vellmer C, Wedepohl KH. Geochemical characterization and origin of granitoids from the south bohemian batholith in Lower Austria. Contributions to Mineralogy and Petrology. 1994;**118**:13-32. DOI: 10.1007/BF00310608

[5] Gerdes A. Geochemische und thermische Modelle zur Frage der spätorogenen Granitgenese am Beispiel des Südböhmischen Batholiths: Basaltischen Underplating oder Krustenstapelung. [Thesis]. Germany: University of Göttingen; 1997

[6] Geological map of the Czech Republic 1:50 000, Czech Geological Survey 2022. Available from: http://www.geology.cz.

[7] Matějka D, Janoušek V. Whole-rock geochemistry and petrogenesis of granites from the northern part of the Moldanubian batholith (Czech Republic). Acta Universitatis Carolineae, Geology. 1998;**42**:73-79

[8] Matějka D, René M. Granitoids of the Freistadt and Mauthausen type in the Czech part of the Moldanubian batholith. In: Pangeo 1. Salzburg: Salzburg University; 2002. pp. 117-118

[9] Mastíková E. Geological Documentation of the Pohled Quarry (Moldanubian Zone). Olomouc, Czech Republic: Palacký University; 2009 [Thesis] (in Czech)

[10] Finger F, Gerdes A, René M, Riegler G. The Saxo-Danubian Belt: Magmatic response to post-collisional delamination of mantle lithosphere below the south-western sector of the bohemian massif (Variscan orogen). Geologica Carpathica. 2009;**60**:205-212. DOI: 10.2478/v10096-009-0014-3

[11] Veselá M et al. Geological Map of the Czech Republic 1:50 000, Sheet 23–23 Jihlava. Prague: Czech Geological Survey; 1991. Available from: http://www.geology.cz

[12] Beneš K et al. Explanations to the Geological Map ČSSR 1:200 000, Sheet M-33-XXIII Jihlava. Prague: Czech Geological Survey; 1963. p. 179 (in Czech)

[13] Doleželová T. Hydrothermal Alterations of Selected Rocks Types in the Havlíčkův Brod Ore District (Moldanubian Zone). Olomouc, Czech Republic: Palacký University; 2006 [Thesis] (in Czech)

[14] Pouchou JL, Pichoir F. PAP (φ-ρ-Z) procedure for improved quantitative

microanalysis. In: Armstrong JT, editor. Microbeam Analysis. San Francisco: San Francisco Press; 1985. pp. 104-106

[15] Richard LR. MinPet: Mineralogical and Petrological Data Processing System, Version 2.02. Québec, Canada: MinPet Geological Software; 1988-1995

[16] Piccoli PM, Candela PA. Apatite in igneous systems. In: Kohn MJ, Rakovan J, Hughes JM, editors. Phosphates – Geochemical, Geobiological and Materials Importance. Vol. 48. Chantilly, Virginia: Reviews in Mineralogy and Geochemistry; 2002. pp. 255-292. DOI: 10.2138/rmg.2002.48.6

[17] Pyle JM, Spear FS, Rudnick RL, McDouough WF. Monazite xenotime-garnet equilibrium in metapelites and a new monazite-garnet thermometer. Journal of Petrology. 2001;**42**:2013-2107. DOI: 10.1093/petrology/42.11.2083

[18] Förster HJ. The chemical composition of REE-Y-Th-U-rich accessory minerals in peraluminous granites of the Erzgebirge-Fichtelgebirge region, Germany. Part I: The monazite-(Ce)-brabantite solid solutions series. American Mineralogy. 1998;**83**:259-272. DOI: 10.2138/am-1998-3-409

[19] René M. Source compositions and melting temperatures of the main granitic suites from the Moldanubian batholith. Journal of Geosciences. 2016; **61**:355-370. DOI: 10.3190/jgeosci.223

[20] Gerdes A. Magma homogenisation during anataxis, ascent and/or emplacement? Constraints from the Variscan Weinsberg granites. Terra Nova. 2001;**13**:305-312. DOI: 10.1046/j.1365-3121.2001.00365.x

[21] Finger F, Clemens J. Migmatization and "secondary" granitic magmas: Effects of emplacement and crystallization of "primary" granitoids in southern bohemian massif, Austria. Contributions to Mineralogy and Petrology. 1995;**120**:311-326. DOI: 10.1007/BF00306510

[22] Sylvester PJ. Post-collisional strongly peraluminous granites. Lithos. 1998;**45**: 29-44. DOI: 10.1016/S0024-4937(98)00024-3

[23] Watson EB, Harrison TM. Zircon saturation revisited: Temperature and composition effects in a variety of crustal magma types. Earth Planetary Science Letters. 1983;**64**:295-304. DOI: 10.1016/0012-821X(83)90211-X

[24] Boehnke P, Watson EB, Trail D, Harrison TM, Schmitt AK. Zircon saturation re-revisited. Chemical Geology. 2013;**351**:324-334. DOI: 10.1016/j.chemgeo.2013.05.028

[25] Montel JM. A model for monazite melt equilibrium and applications to generation of granitic magmas. Chemical Geology. 1993;**110**:127-146. DOI: 10.1016/0009-2541(93)90250-M

[26] Miller CF. Hot and cold granites? Implications of zircon saturation temperatures and preservation of inheritance. Geology. 2003;**31**:529-532. DOI: 10.1130/0091-7613(2003)031

[27] Siégel C, Bryan SE, Allen CM, Gust DA. Use and abuse of zircon-based thermometers: A critical review and a recommended approach to identify antecrystic zircon. Earth-Science Reviews. 2018;**176**:87-116. DOI: 10.1016/j.earscirev.2017.08.011

[28] Clemens JD, Stevens G, Bryan SE. Conditions during the formation of granitic magmas by crustal melting – Hot or cold; drenched, damp or dry? Earth-Science Reviews. 2020;**200**:1-21. DOI: 10.1016/j.earscirev.2019.102982

[29] René M. Occurrence of Th, U, Y, Zr and REE-bearing accessory minerals in granites and their petrogenetic significance. In: Blasik M, Bogdashka H, editors. Granite Occurrence, Mineralogy and Origin. New York: Nova Science Publishers, Inc.; 2012. pp. 27-56

[30] René M. REE and Y mineralogy of the Moldanubian batholith (central European Variscides). In: Veress B, Szigethy J, editors. Horizons in Earth Science Research. New York: Nova Science Publishers, Inc.; 2017. pp. 73-114

Chapter 3

Fluorocarbonates from Biotite Granodiorite, Slavkovský les Mts., Czech Republic

Miloš René and Zdeněk Dolníček

Abstract

Biotite granodiorites belonging to the redwitzite suite of western part of the Bohemian Massif occur as small bodies in metasediments of the Horní Slavkov crystalline unit and/or as inclusions in biotite granites of the Krušné Hory/Erzgebirge Mts. batholith. Biotite granodiorites contain plagioclase, K-feldspar, biotite, quartz, and accessory minerals (apatite, zircon). Some of these granodiorites were hydrothermally altered and during breakdown of biotite originated chlorite, titanite, ilmenite, and REE-fluorocarbonates. The anhedral grains and irregular aggregates of REE-fluorocarbonates appear homogenous in composition in back-scattered electron images. However, the detailed microprobe analyses of individual REE-fluorocarbonate grains show relatively high compositional variability on thin section scale, particularly with respect to their contents of Ca and Y. The REE-fluorocarbonates are represented by parisite, bastnäsite and relatively rare synchysite. REE-fluorocarbonates are more widespread in the Variscan granites of the Krušné Hory/Erzgebirge Mts. and Slavkovský les Mts. than previously considered. The occurrence of these REE-fluorocarbonates demonstrates that during later post-magmatic alterations, primary accessory minerals (allanite, monazite, xenotime, zircon) became unstable with remobilization of REE, Th, and U into newly formed secondary minerals (REE-fluorocarbonates and coffinite).

Keywords: parisite, bastnäsite, synchysite, redwitzite, chloritization, geochemistry, bohemian massif, central European variscides

1. Introduction

The bulk of the rare earth elements (REE) in granitic rocks is contained in primary accessory minerals, such as allanite, monazite, apatite, xenotime, and zircon. During post-magmatic hydrothermal alteration, biotite is usually altered to chlorite.

Hydrothermal alteration-related REE minerals can be used as useful tracers of fluid composition and crystallization conditions. Normalized REE patterns of hydrothermal minerals are only rarely directly reflecting those of source rocks due to usually strong crystallochemical control and chemical complexation of REEs. Because of changing the affinity of various REEs to complexing agents present in the fluid phase, various levels of fractionation of REEs are often observed in hydrothermal REE minerals, especially in rare-earth fluorocarbonates.

The original, magmatically crystallized accessories may be corroded, or even completely dissolved. Under the influence of fluorine- and CO_2-bearing fluids, REE become preferentially redeposited as rare-earth fluorocarbonates, such as parasite, bastnäsite, and synchysite. These REE-fluorocarbonates occur in various geological environments, including carbonatites, alkaline magmatic rocks, weathering crusts, sedimentary rocks, as well as hydrothermal ore deposits [1–21]. The most abundant REE-fluorocarbonate minerals are bastnäsite and synchysite, which have also economic importance as one of the major world supply sources of REE (Mountain Pass, California, USA; Bayan Obo, Inner Mongolia, China).

Parisite is a highly scarce representative of this mineral group and in the past, it was described from altered granites and acid volcanites [7, 11, 17, 18], carbonatite complexes [9, 11], and hydrothermal ore deposits [3, 4, 8, 18]. In altered granitic rocks of the Saxothuringian Zone, the most abundant REE-fluorocarbonates were bastnäsite and synchysite [3, 10, 12, 22].

The present paper describes the occurrence and detailed mineralogy of REE-fluorocarbonates (parisite, bastnäsite, and rare synchysite) in altered granodiorites (redwitzites) of the Krudum granite body that is an integral part of the Western Krušné Hory/Erzgebirge pluton [23–25].

2. Geological setting

Biotite granodiorites are part of the oldest redwitzite suite that occurred in the area of the Krudum granite body (KGB), which is a subsidiary intrusion of the Karlovy Vary pluton (KVP) (**Figure 1**) [25]. Redwitzite suite that was originally defined in

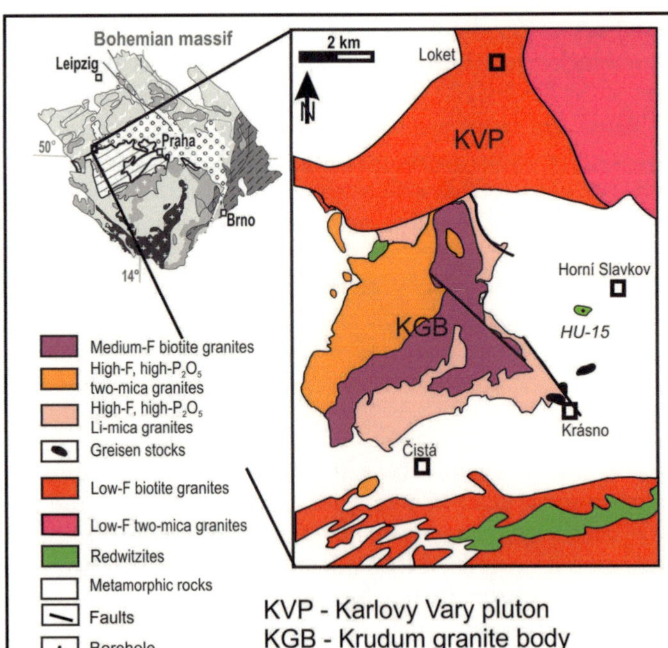

Figure 1.
Geological map of the Krudum granite body (after [25] modified by authors).

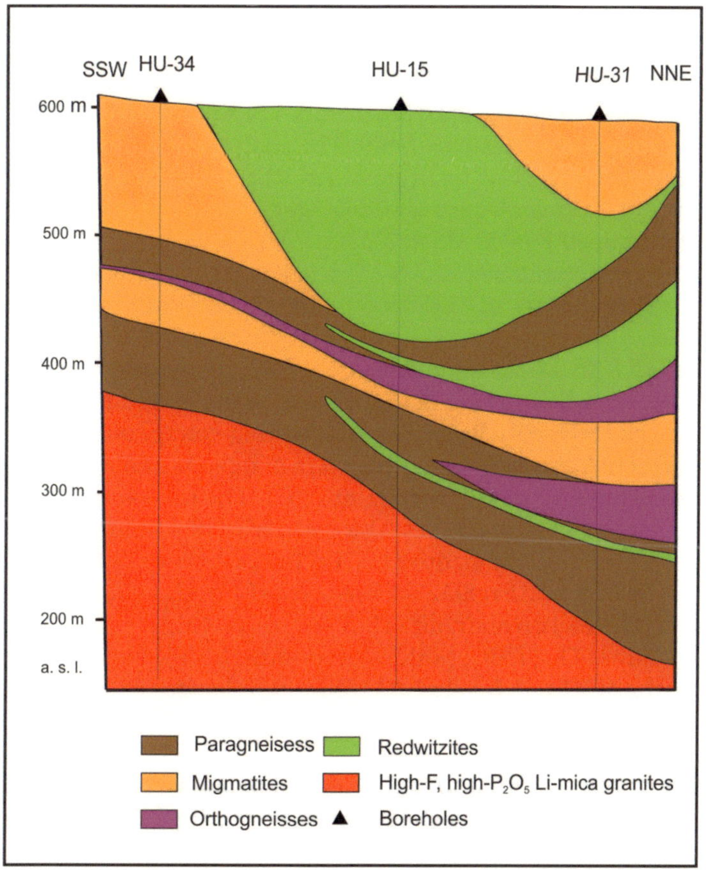

Figure 2.
Geological cross section of the redwitzite bodies from the Slavkovský les Mts. (after [31] modified by authors).

NE Bavaria comprises a small group of mafic to intermediate magmatic rocks ranging from gabbros to granodiorites [26–28]. Later, rocks similar to redwitzites were described from other Variscan plutons of the north-western edge of the Bohemian Massif [29, 30].

During exploration of the Sn-W mineralization in the Horní Slavkov-Krásno ore district the rocks of redwitzite suite were found as irregular magmatic bodies and sills in metamorphic rocks of the Slavkov crystalline unit (**Figure 2**) [31]. The granitic rocks of the Krudum granite body intrude the Slavkov crystalline unit composed of paragneisses, migmatites, and orthogneisses. The granitoid sequence of this body consists of a number of intrusive phases from the oldest redwitzite suite to the youngest F-rich, high-P_2O_5 Li-mica granites connected with Sn-W greisen mineralization occurred in greisen stocks (e.g., Hub stock) [32].

3. Materials and methods

Two representative rock samples of biotite granodiorites (redwitzites) were taken from borehole HU-15, which was performed by SW of the Horní Slavkov town (**Figure 2**). The borehole encountered three granodiorite sills in paragneisses and migmatites of the Slavkov crystalline unit above hidden part of the F-rich, high-P_2O_5 Li-mica granites. Two representative rocks samples of redwitzites weighing 1 kg were crushed in a jaw crusher and representative splits of these samples were ground in an agate ball mill. The analyses of major elements were performed in the laboratory of Institute of Rock Structure and Mechanics of the Czech Academy of Sciences by conventional wet chemical methods. Trace elements were determined by ICP MS at Activation Laboratories Ltd., Lancaster, Canada on a Perkin Elmer Sciex ELAN 6100 ICP mass spectrometer. The decomposition of rock samples for ICP MS analyses involved lithium metaborate/tetraborate fusion.

Analyses of all REE-fluorocarbonates were performed using a CAMECA SX 100 electron microprobe working in WDX mode at the Institute of Geological Sciences, Masaryk University Brno. The accelerating voltage and beam currents were 15 kV and 10 nA, respectively, with beam diameters of 8 μm. The raw data were corrected using PAP matrix corrections [33]. The following standards, X-ray lines and crystals (in parentheses) were used: AlK_α—almandine (TAP), AsL_α—InAs (TAP), CaK_α—brabantite (PET), BaL_α—barite (LPET), CeL_α—$CeAl_2$ (PET), DyL_α—$DyPO_4$ (LLIF), ErL_α—YErAG (LLIF), EuL_β—$EuPO_4$ (LLIF), FK_α—topaz (PC1), FeK_α on andradite (LLIF), GdL_β—GdF_3 (LLIF), LaL_α—LaB_6 (PET), MgK_α—$MgAl_2O_4$ (TAP), MnK_α—rhodonite (LLIF), NdL_β—NdF_3 (LLIF), PK_α—fluorapatite (PET), PrL_β—PrF_3 (LLIF), SK_α—barite (PET), SiK_α—spessartine (TAP), SmL_β—SmF_3 (LLIF), SrL_α—$SrSO_4$ (TAP), ThM_α—brabantite (LPET), UM_β—metallic uranium (PET), YL_α—YAG (PET), and ZrL_α—zircon (TAP). Elements were measured for 20 s at the peak and for 10 s for each background. The detection limits were approximately 100 ppm for Y, 180–1700 ppm for REE, and 800–1100 ppm for U and Th.

4. Results

4.1 Petrology

The analyzed medium-grained biotite granodiorites consist of euhedral quartz (18–20 vol.%), anhedral to subhedral plagioclase (An_{27-37}) (30–40 vol.%), andehral to subhedral K-feldspar (15–20 vol.%), and anhedral biotite (10–15 vol.%) (**Figure 3**). The unaltered biotite is characterized by distinctly higher content of TiO_2 (3.8–4.6 wt.%) and usually lower concentrations of F (0.09–0.42 wt.% F). Biotite is represented by annite (Fe/(Fe + Mg) = 0.56–0.59, ^{IV}Al = 2.02–2.32 apfu and Ti = 0.47–0.59 apfu). Biotite often contains inclusions of accessories (apatite, zircon) and is partly altered to a mixture of chlorite, titanite, ilmenite, and REE-fluorocarbonates. Sericitization of both feldspars and chloritization of biotite are widespread, especially in sample R-1359.

The anhedral grains and irregular REE-fluorocarbonate aggregates appear homogenous in back-scattered electron images (**Figures 4** and **5**). However, the detailed microprobe analyses of individual REE-fluorocarbonate grains and aggregates show relatively high compositional variability on thin section scale, particularly with respect to their contents of Ca and Y (**Table 1**).

Figure 3.
Microphotograph of the biotite granodiorite (redwitzite) (Bt—biotite, Kfs—K-feldspar, Pl—plagioclase, Qz—quartz), thin section, crossed polarizers.

Figure 4.
Back-scattered electron (BSE) image of REE-fluorocarbonates from altered redwitzite (Ap—apatite, Prs—parisite).

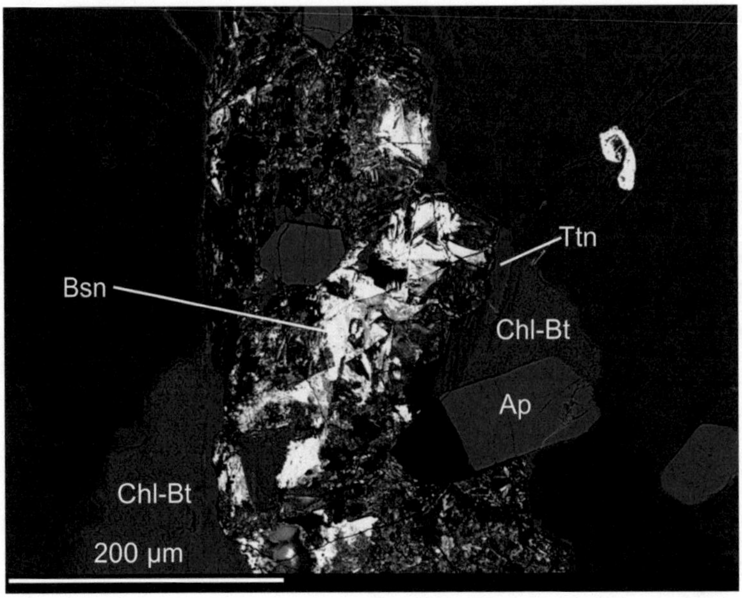

Figure 5.
Back scattered electron (BSE) image of REE-fluorocarbonates from altered redwitzite (Ap—apatite, Bsn—bastnäsite, Chl-Bt—chloritized biotite, Ttn–titanite).

Sample wt.%	R-1359	R-1360
SiO_2	57.10	57.41
TiO_2	1.58	1.65
Al_2O_3	17.16	17.11
Fe_2O_3	1.88	1.73
FeO	5.02	5.04
MnO	0.09	0.09
MgO	2.61	2.57
CaO	4.13	4.59
Na_2O	3.10	3.33
K_2O	3.31	2.99
P_2O_5	0.55	0.48
H_2O^+	1.86	1.57
H_2O^-	0.43	0.31
Total	98.82	98.87
A/CNK	1.06	1.00
Ba ppm	2100	2110
Sr	734	764
Y	22	20
Zr	322	430

Table 1.
Representative compositions of the Horní Slavkov redwitzites.

4.2 Geochemistry

Analyzed biotite granodiorites are weakly peraluminous rocks, which show I-type signature with A/CNK (mol. $Al_2O_3/(CaO + Na_2O + K_2O)$ of 1.00–1.06 and contents of SiO_2 56.3–57.1 wt.%, TiO_2 1.58–1.70 wt.% and MgO 2.60–2.61 wt.%. Potassium concentrations are higher (2.77–3.31 wt.% K_2O) and K_2O/Na_2O ranges between 0.86 and 1.07, showing that these granodiorites belong to high-K magmas [34]. In addition, high content of Ba (1939–2000 ppm), Sr (680–715 ppm), and Zr (256–315 ppm) are significant (**Table 2**). The REE patterns of biotite granodiorites normalized to chondrite show the predominance of LREE over HREE (La_N/Yb_N = 40.17–52.20) and gently negative Eu anomaly (Eu/Eu^* = 0.85–0.87) (**Figure 6**).

wt.%	1a	2a	3a	4	8a	10	10-1
SO_3	b.d.l.	b.d.l.	b.d.l.	0.01	b.d.l.	b.d.l.	b.d.l.
P_2O_5	0.04	0.01	0.01	0.05	0.14	0.04	0.02
As_2O_5	0.07	0.07	b.d.l.	b.d.l.	b.d.l.	0.03	b.d.l.
CO_2	22.72	21.98	22.07	21.35	20.93	22.24	25.34
SiO_2	b.d.l.	0.21	0.15	b.d.l.	3.16	1.95	b.d.l.
ThO_2	2.50	1.94	1.83	2.21	3.25	4.59	0.46
UO_2	0.32	0.30	0.29	0.12	0.27	0.24	0.04
ZrO_2	0.04	b.d.l.	b.d.l.	b.d.l.	b.d.l.	b.d.l.	b.d.l.
Al_2O_3	0.00	0.02	0.03	b.d.l.	1.61	0.97	b.d.l.
Y_2O_3	0.51	0.55	0.51	2.03	1.00	0.99	0.77
La_2O_3	16.78	17.62	17.76	13.53	15.06	15.07	11.75
Ce_2O_3	31.56	31.62	31.84	27.09	28.55	29.48	23.48
Pr_2O_3	3.04	2.99	3.07	2.79	2.96	2.81	2.53
Nd_2O_3	10.27	10.41	10.31	10.28	10.04	10.14	9.17
Sm_2O_3	1.19	1.27	1.25	1.52	1.37	1.29	1.27
Eu_2O_3	b.d.l.	b.d.l.	b.d.l.	0.01	b.d.l.	b.d.l.	b.d.l.
Gd_2O_3	0.37	0.41	0.41	0.94	0.52	0.62	0.50
Dy_2O_3	b.d.l.	0.12	0.09	0.46	0.28	0.03	0.09
Er_2O_3	b.d.l.	0.02	b.d.l.	0.20	0.06	0.13	0.04
FeO	1.04	0.51	0.33	0.29	0.22	0.18	1.52
MgO	0.05	0.05	0.03	0.18	0.13	0.10	0.45
CaO	5.43	4.73	4.86	5.84	6.21	6.80	13.28
SrO	0.60	0.62	0.63	0.28	0.29	0.37	0.25
PbO	0.02	0.03	b.d.l.	0.02	0.03	0.01	b.d.l.
H_2O^*	0.47	1.81	1.83	0.17	0.28	0.43	0.14
F	5.56	5.76	5.73	5.80	6.13	5.91	5.18
O=F	−2.34	−2.56	−2.77	−2.44	−2.58	−2.49	−2.56
Total	100.24	100.47	100.26	92.71	99.91	102.18	93.72

wt.%	1a	2a	3a	4	8a	10	10-1
S^{6+}	0.000	0.000	0.000	0.000	0.000	0.000	0.000
P^{5+}	0.004	0.000	0.000	0.004	0.011	0.001	0.001
As^{5+}	0.003	0.001	0.000	0.000	0.000	0.000	0.000
C^{4+}	2.993	0.992	0.995	2.996	2.691	1.999	1.999
Si^{4+}	0.000	0.007	0.005	0.000	0.298	0.000	0.000
Subtotal	3.000	1.000	1.000	3.000	3.000	2.000	2.000
Th^{4+}	0.055	0.015	0.014	0.052	0.070	0.006	0.006
U^{4+}	0.007	0.002	0.002	0.003	0.006	0.001	0.001
Zr^{4+}	0.002	0.000	0.000	0.000	0.000	0.000	0.000
Al^{3+}	0.000	0.001	0.001	0.000	0.178	0.000	0.000
Y^{3+}	0.026	0.010	0.009	0.111	0.050	0.024	0.024
La^{3+}	0.597	0.215	0.216	0.513	0.523	0.250	0.250
Ce^{3+}	1.115	0.383	0.385	1.019	0.984	0.497	0.497
Pr^{3+}	0.107	0.036	0.037	0.104	0.102	0.053	0.053
Nd^{3+}	0.354	0.123	0.122	0.377	0.338	0.189	0.189
Sm^{3+}	0.040	0.014	0.014	0.054	0.044	0.025	0.025
Eu^{3+}	0.000	0.000	0.000	0.000	0.000	0.000	0.000
Gd^{3+}	0.012	0.004	0.004	0.032	0.016	0.010	0.010
Dy^{3+}	0.000	0.001	0.001	0.015	0.009	0.002	0.002
Er^{3+}	0.000	0.000	0.000	0.006	0.002	0.001	0.001
Fe^{2+}	0.084	0.014	0.009	0.025	0.017	0.074	0.074
Mg^{2+}	0.007	0.003	0.002	0.027	0.019	0.039	0.039
Ca^{2+}	0.561	0.168	0.172	0.644	0.627	0.822	0.822
Sr^{2+}	0.034	0.012	0.012	0.017	0.016	0.008	0.008
Pb^{2+}	0.000	0.000	0.000	0.000	0.001	0.000	0.000
Na^{+}	0.000	0.000	0.000	0.000	0.000	0.000	0.000
Subtotal	3.001	1.000	1.000	3.000	3.000	2.000	2.000
OH^{-*}	0.302	0.399	0.403	0.117	0.176	0.054	0.054
F^{-}	1.697	0.602	0.598	1.885	1.826	0.946	0.946
Subtotal	2.000	1.001	1.001	2.002	2.002	1.000	1.000
O^{2-}	10.344	3.610	3.612	10.232	10.292	10.324	6.559
La/Ce	0.53	0.56	0.56	0.50	0.53	0.51	0.51
La/Nd	1.63	1.69	1.72	1.62	1.50	1.49	149
Mineral	Parisite	bastnäsite	bastnäsite	parisite	parisite	parisite	Synchysite

b.d.l.—below detection limit.

Table 2.
Representative analyses of fluorocarbonates from the altered redwitzites of the Krudum granite body.

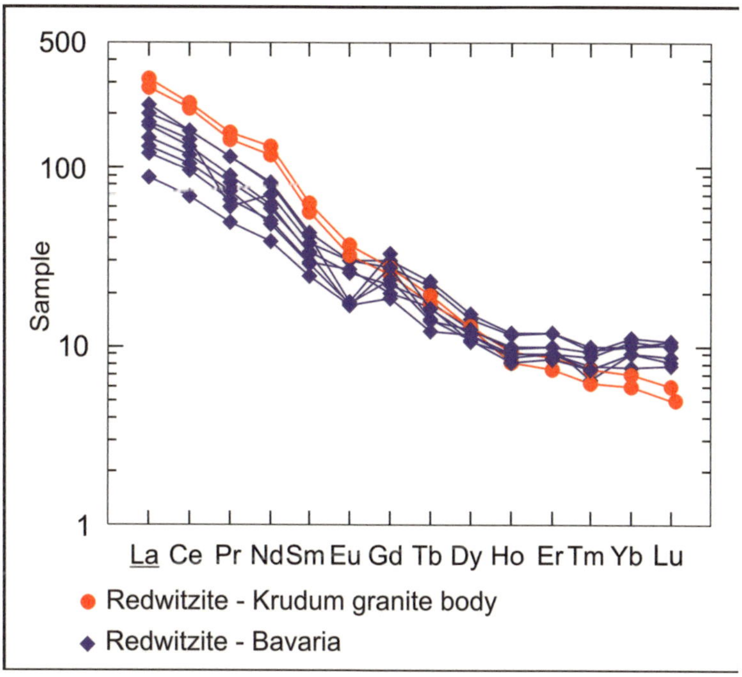

Figure 6.
Chondrite normalized REE diagram of redwitzites from the Krudum granite body (this work) and Bavaria (data from [34]) by [35].

4.3 Fluorocarbonate composition

The REE-fluorocarbonates have Ce as the dominant REE cation, and should be named as parisite-(Ce), bastnäsite-(Ce), and/or synchysite-(Ce). The concentration of the most abundant LREE oxide, Ce_2O_3, decreases from 28.2 to 26.5 wt.%. The thorium is present in concentrations between 2.0 and 4.2 wt.% of ThO_2. The content of yttrium is considerably low (0.8–1.1 wt.% Y_2O_3) (**Figure 7**). Relatively low in all analyzed REE fluorocarbonates are also concentrations of neodymium (2.6–10.7 wt.% Nd_2O_3) (**Figure 8**).

5. Discussion

5.1 Petrology and geochemistry of redwitzites

Redwitzites were originally found in the NW part of the Bohemian massif in NE Bavaria (Germany) [26, 27] and comprise mafic to intermediate Variscan igneous rocks ranging from gabbros to granodiorites. These rocks represent very specific rock type which represent the oldest rock variety that evolved during the Variscan magmatic phase in the north-west part of the Bohemian Massif [28, 34]. Later were similar rocks found and described also from the area of the Krudum granite body [23, 24, 29]. In area of the Krudum granite body occur monzodiorites, quartz diorites and granodiorites composed of plagioclase (An_{24-52}) (30–50 vol.%), K-feldspar (5–20 vol.%), quartz (5–20 vol.%), biotite (10–30 vol.%), hornblende (0–24 vol.%), and rare pyroxene (0–5

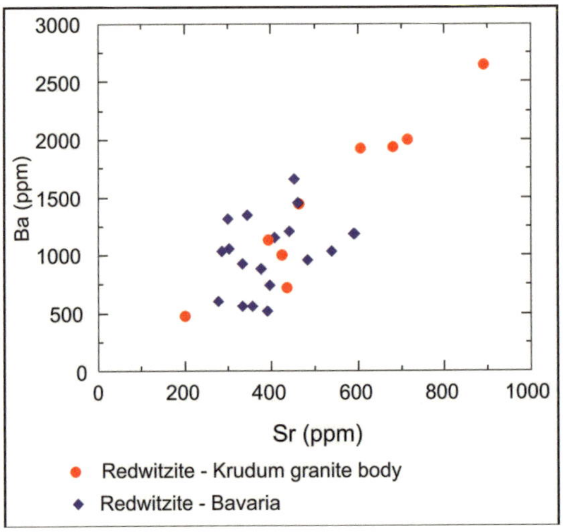

Figure 7.
Distribution of La, Ce and Y in REE-fluorocarbonates from redwitzites of the Krudum granite body (this work) and from Li-mica granites of the Krušné Hory/Erzgebirge Mts. (data taken from [3, 22]).

vol.%). Composition of biotite range from phlogopite to annite. For biotite from the Bavarian redwitzites the presence of phlogopite is significant (Fe/(Fe + Mg) = 0.46–0.51, ^{IV}Al = 2.30–2.40 apfu and Ti = 0.32–0.53 apfu). Accessory minerals are apatite, zircon, ilmenite, titanite, allanite, rutile, magnetite, and pyrrhotite [23, 24].

The SiO_2 contents in the Bavarian redwitzites fluctuates over a wide range (49–63 wt.%). The TiO_2 content in the Bavarian redwitzites is partly lower as its

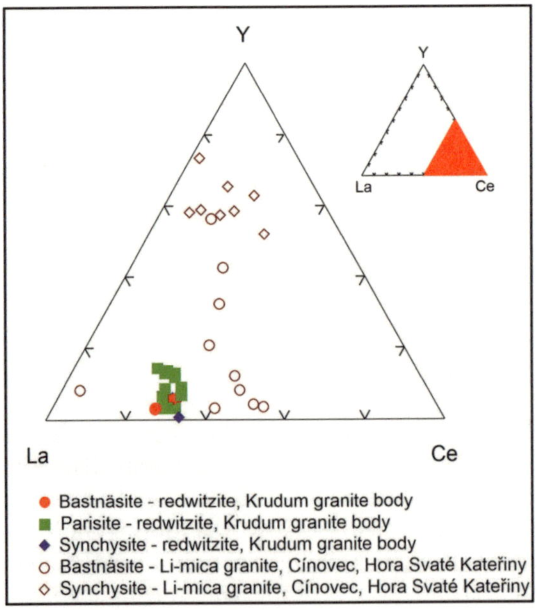

Figure 8.
Distribution of La, Ce and Nd in REE-fluorocarbonates from redwitzites of the Krudum granite body (this work) and from Li-mica granites of the Krušné Hory/Erzgebirge Mts. (data taken from [3, 22]).

content in redwitzites from the Krudum granite body (0.69–1.88 wt.%). Potassium concentrations in the Bavarian redwitzites are partly higher than the concentrations in redwitzites from the Slavkov granite body (1.40–4.37 wt.% K_2O) and K_2O/Na_2O ranges between 0.53 and 1.49. The analyzed Bavarian redwitzites display lower contents of Ba (513–1656 ppm), Sr (278–591 ppm), and Zr (157–371 ppm) than the similar rocks from the Krudum granite body. The REE pattern of biotite granodiorites normalized to chondrite shows the predominance of LREE over HREE (La_N/Yb_N = 8.27–24.48) and partly negative Eu anomaly (Eu/Eu^* = 0.47–0.95) (**Figure 6**) [34].

5.2 Occurrence of REE-fluorocarbonates in the Krušné Hory Mts./Erzgebirge granitic rocks

Rare-earth fluorocarbonates are relatively widespread in some hydrothermally altered Variscan granites of the Krušné Hory/Erzgebirge Mts. and Slavkovský les Mts. Occurrence of these REE-fluorocarbonates demonstrates that during later post-magmatic alterations primary accessory minerals (allanite, monazite, xenotime, zircon) became unstable with the remobilization of REE, Th, and U into newly formed secondary minerals (REE-fluorocarbonates, highly rare REE-arsenates and coffinite). In addition to newly founded REE-fluorocarbonates from the Krudum granite body, REE-fluorocarbonates and rare REE-arsenates were noted from the Niederbobritzsch granite (cerite-(Ce), synchysite-(Ce) [10], Kirchberg granite (bastnäsite-(Ce), Altenberg granite (fluorocerite-(Ce), bastnäsite-(Ce) [36], Cínovec granite (bastnäsite-(Ce), synchysite-(Ce), synchysite-(Y), chernovite-(Y) [3], Hora Svaté Kateřiny granite (bastnäsite-(Ce), bastnäsite-(Y), chernovite-(Y) [22], Gottesberg microgranite (synchysite-(Ce) [37] and Markersbach granite (synchysite-(Ce), synchysite-(Y) [12]. For all these occurrences is significant the predominance of bastnäsite-(Ce) and synchysite-(Ce). The La/Ce ratio for bastnäsite-(Ce) from hydrothermally altered, low-P_2O_5, high-F, Li-mica granites from the Krušné Hory/Erzgebirge batholith (Cínovec, Markersbach, Hora Svaté Kateřiny [3, 12, 22] is 0.26–0.83 and La/Nd ratio for bastnäsite-(Ce) from these rocks is 0.45–1.67. For synchysite-(Ce) from these rocks, the La/Ce ratio is 0.13–0.46 and the La/Nd ratio is 0.53–1.23. In REE-fluorocarbonates from redwitzites of the Krudum granite body are in comparison with REE-fluorocarbonates from above mentioned Li-mica granites distinctly lower concentrations of yttrium and also partly lower the concentrations of neodymium (**Figures** 7 and **8**).

5.3 Chemical composition and origin of REE-fluorocarbonates

Although REE-fluorocarbonates have been described from several carbonatite complexes and hydrothermally altered granitic rocks, detailed compositional data are relatively rare. Identification of these minerals is commonly based on microprobe chemical analyses.

Our contribution is concentrated on occurrences REE-fluorocarbonates in hydrothermal altered redwitzites, which represent a highly rare occurrence of these minerals in granitic and related rock types, not yet published from other parts of the world. In previous contributions concentrated on occurrence of REE-fluorocarbonates in hydrothermal altered rocks, the older studies were concentrated on hydrothermal altered rocks coupled with uranium mineralization [4, 18] and on hydrothermal alteration of highly fractionated granitic rocks [3, 10, 12, 20, 22].

The majority of previously examined parisites-(Ce) contain Th concentrations below 1 wt.%. The Th-enriched parisite-(Ce) was found in carbonatites from Malawi and Brazil [15, 16]. The occurrence of REE-fluorocarbonates requires the presence of fluorine- and carbon dioxide-bearing fluids. The crystallization of parisite instead of bastnäsite indicates that the activity of Ca^{2+} in that fluid was distinctly higher. The primary magmatic plagioclase has a sufficient amount of calcium, which could be liberated during its sericitization. Potential source of fluorine is biotite, however, its content in biotite from examined granodiorites is relatively low (0.09–0.18 wt.% F). At the Horní Slavkov-Krásno Sn-W ore district carbonate country rocks are absent. On the other hand, CO_2 has been proven in the residual melt fractions of Li-mica granites [38]. Moreover, the CO_2-enriched fluid inclusions were recently found in quartz from Sn-W mineralization of the Horní Slavkov-Krásno ore district [39].

According to Fleischer [40] the average La/Nd ratio for REE-fluorocarbonates from hydrothermally altered granitic rocks is 0.89–1.83. Similar value gives for synchysite from hydrothermally altered granitic rocks also Schmandt et al. [18]. These values well correspond with La/Nd ratios which were found for parisite (1.33–1.63) and bastnäsite (1.56–1.72) from examined hydrothermally altered redwitzites of the Krudum granite body. The evolution of La/Ce ratio in bastnäsite from different geological environments was studied by Schmandt et al. [18] and for bastnäsite from hydrothermally altered granitic rocks, they give a value of 0.45–0.50. These data could very good correlate with our data for bastnäsite (0.49–0.56) and parisite (0.47–0.53).

6. Conclusions

The analyzed medium-grained biotite granodiorites from the Krudum granite body consist of quartz, plagioclase (An_{27-37}), K-feldspar, and biotite. The unaltered biotite is characterized by the distinctly higher content of TiO_2. Biotite is represented by annite (Fe/(Fe + Mg) = 0.56–0.59). Biotite often contains inclusions of accessories (apatite, zircon) and is partly hydrothermally altered to a mixture of chlorite, titanite, ilmenite, and REE-fluorocarbonates. Analyzed unaltered biotite granodiorites are weakly peraluminous rocks, which show an I-type signature.

The anhedral grains and irregular aggregates of REE-fluorocarbonates appear homogenous in back-scattered electron images. However, the detailed microprobe analyses of individual REE-fluorocarbonate grains and aggregates show relatively high compositional variability on a thin section scale, particularly with respect to their contents of Ca and Y. The REE-fluorocarbonates have Ce as the dominant REE cation, and should be named as parisite-(Ce), bastnäsite-(Ce), and/or very rare synchysite-(Ce). However, for analyzed REE-fluorocarbonates from biotite granodiorites (redwitzites) are considerably low concentrations of yttrium (0.8–1.1 wt.% Y_2O_3).

The La/Nd ratios which were found for parisite (1.33–1.63) and bastnäsite (1.56–1.72), occurred in redwitzites from the Krudum granite body are characteristic of REE-fluorocarbonates from hydrothermally altered granitic rocks. Also, the La/Ce ratios in analyzed bastnäsite (0.49–0.56) and parisite (0.47–0.53) are similar to these ratios which were found in REE-fluorocarbonates from hydrothermally altered granitic rocks.

Acknowledgements

The support of the Long-Term Conceptual Development Research Organization RVO 67985891 is thanked, Z.D. acknowledges the financial support of the Ministry of

Culture of the Czech Republic (long-term project DKRVO 2019–2023/1.i.e; National Museum, 00023272). We are also grateful to Š. Benedová and R. Škoda for their technical assistance in using electron microprobe analyses of REE-fluorocarbonates, plagioclase, and biotite. We are also grateful for the constructive comments of an anonymous reviewer.

Conflict of interest

The authors declare no conflict of interest.

Author details

Miloš René[1]* and Zdeněk Dolníček[2]

1 Institute of Rock Structure and Mechanics, v.v.i., Czech Academy of Sciences, Prague, Czech Republic

2 Department of Mineralogy and Petrology, National Museum, Prague, Czech Republic

*Address all correspondence to: rene@irsm.cas.cz

References

[1] Littlejohn AI. Alteration products of accessory allanite in radioactive granites from the Canadian Shield. Geological Survey of Canada Paper. 1981;**81-1B**:95-104

[2] Williams-Jones AE, Wood SA. A preliminary petrogenetic grid for REE fluorocarbonates and associated minerals. Geochimica et Cosmochimica Acta. 1992;**56**:725-738. DOI: 10,1016/0016-7037(92)90093-X

[3] Johan Z, Johan V. Accessory minerals of the Cínovec (Zinnwald) granite cupola, Czech Republic: Indicators of petrogenetic evolution. Mineralogy and Petrolology. 1995;**83**:113-150. DOI: 10.1007/s00710-004-0058-0

[4] Lottermoser BG. Rare earth element mineralogy of the olympic dam Cu-U-Au-Ag deposit, Roxby Downs, South Australia: Implications for ore genesis. Neues Jahrbuch für Mineralogie Monatshefte. 1995;**1995**(8):371-384

[5] Gieré R. Formation of rare earth minerals in hydrothermal systems. In: Jones AP, Wall F, Williams CT, editors. Rare Earth Minerals: Chemistry, Origin and Ore Minerals. London: Chapman and Hall; 1996. pp. 105-150

[6] de Parseval P, Fontan F, Aigouy T. Chemical composition of rare earth minerals from Trimouns Ariége, France. Comptes Rendus de l´ Académie des Sciences Serie II. 1997;**324**:625-630

[7] Pan Y. Zircon- and monazite forming metamorphic reactions at Manitouwadge, Ontario. Canadian Mineralogist. 1997;**35**:105-118

[8] Iida Y, Ohnuki T, Isobe H, Yanase N, Sekine K, Yoshiba H, et al. Hydrothermal redistribution of rare earth elements in Toki granitic rock, central Japan. Journal of Contaminant Hydrology. 1998;**35**:191-199. DOI: 10.1016/50169-7722(98)00130-2

[9] Zaitsev AN, Wall F, LeBas MJ. Sr-Ba minerals from the Khibina Carbonatites, Kola Peninsula, Russia: Their mineralogy, paragenesis and evolution. Mineralogical Magazine. 1998;**62**:225-250. DOI: 10.1180/002646198547594

[10] Förster HJ. Cerite-(Ce) and thorian synchysite from Niederbobritzsch (Erzgebirge, Germany): Implications for differential mobility of Th and the LREE during granite alteration. Canadian Mineralogist. 2000;**38**:67-79. DOI: 10.2113/gscanmin.38.1.67

[11] Smith MP, Henderson P, Campbell LS. Fractionation of the REE during hydrothermal processes. Constrains from the Bayan Obo Fe-REE-Nb deposit, Inner Mongolia, China. Geochimica et Cosmochimica Acta. 2000;**64**:3141-3160. DOI: 1016/50016-7037(00)00416-6

[12] Förster HJ. Synchysite-(Y)-synchysite-(Ce) solid solutions from Markersbach, Erzgebirge, Germany: REE and Th mobility during high-T alteration of highly fractionated aluminous A-type granites. Mineralogy and Petrology. 2001;**72**:259-280. DOI: 10.1007/s007100170019

[13] Ondrejka M, Uher P, Pršek J, Ozdin D. Arsenian monzonite-(Ce) and xenotime-(Y), REE arsenates and carbonates from the Tisovec-Rejkovo rhyolite, West Carpathians, Slovakia: Composition and substitutions in the (REE, Y)XO_4 system (X = P, As, Si, Nb, S). Lithos. 2007;**95**:116-129. DOI: 10.1016/j.lithos.2006.07.019

[14] Holtstam D, Anderson UB. The REE minerals of the Bastnäs-type deposits, South Central, Sweden. Canadian Mineralogist. 2007;**45**:1073-1114. DOI: 10.2113/gscanmin.45.5.1073

[15] Ruberti E, Enrich GER, Gomes CB. Hydrothermal REE fluorcarbonate mineralization at Barra do Itapirapuã, a multiple stockwork carbpnatite, Southern Brazil. Canadian Mineralogist. 2008;**46**:901-914. DOI: 10.374/canmin.46.4.901

[16] Guastoni A, Nestola F, Giaretta A. Mineral chemistry and alteration of rare earth element (REE) carbonatite from alkaline pegmatites of Mount Malosa, Malawi. American Mineralogist. 2009;**94**:1216-1222. DOI: 10.2138/am.2009.3185

[17] Augé T, Bailly L, Wille G. An unusual occurrence of synchysite-(Ce) in amygdules from the Esterel volcanic rocks, France. Implications for rare-earth element mobility. Canadian Mineralogist. 2014;**52**:837-856. DOI: 10.3749/canmin.1400012

[18] Schmandt DS, Cook NJ, Globanu CJ, Ehrig K, Wade BP, Gilbert S, et al. Rare earth element fluorocarbonate minerals from the Olympic Dam Cu-U-Ag-Au deposit, South Australia. Minerals. 2017;**7**:1-24. DOI: 10.3390/min7100202

[19] Caruso L, Simmons G. Uranium and microcracks in a 1000-meter core, Redstone, New Hampshire. Contributions to Mineralogy and Petrology. 1985;**90**:1-17. DOI: 10.1007/BF00373036

[20] Maruéjol P, Cuney M, Turpin L. Magmatic and hydrothermal R.E.E. fractionation in the Xihuashan granites (SE China). Contributions to Mineralogy and Petrology. 1990;**104**:668-680. DOI: 10.1007/BF01167286

[21] Pan Y, Fleet MF, Barnett RI. Rare-earth mineralogy and geochemistry of the Mattagami lake volcanogenic massive sulfide deposit, Quebec. Canadian Mineralogist. 1994;**32**:133-147

[22] Breiter K, Čopjaková R, Škoda R. The involvement of F, CO_2, and As in the alteration of Zr–Th–REE-bearing accessory minerals in the Hora Svaté Kateřiny A-type granite, Czech Republic. Canadian Mineralogist. 2009;**47**:1375-1398. DOI: 10.3749/canmin.47.6.1375

[23] Kovářiková P, Siebel W, Jelínek E, Štemprok M, Kachlík V, Holub FV, et al. Diorite intrusions of the Slavkovský les (Kaiserwald), Western Bohemia: Their origin and significance in late Variscan granitoid magmatism. International Journal of Earth Science. 2010;**99**:545-565. DOI: 10.1007/s00531-008-0406-0

[24] René M. Granodiorites of the redwitzite suite from the drill hole HU-15 in the Horní Slavkov area. Geoscience Research Reports. 2010;**43**:245-247 (In Czech)

[25] Machek M, Roxerová Z, Janoušek V, Staněk M, Petrovský E, René M. Petrophysical and geochemical constraints on alteration processes in granites. Studia of Geophysics and Geology. 2013;**57**:710-740. DOI: 10.1007/s11200-013-0923-6

[26] Willman K. Die Redwitzite, eine neue Gruppe von granitischen Lamprophyren. Zeitschrift der Deutschen Geologischen Gesellschaft. 1920;**71**:1-33

[27] Troll G. Gliederung der redwitzitischen Gesteine Bayerns nach Stoff- und Gefügemerkmalen. Teil 1 Die Typlokalität von Marktredwitz in Oberfranken. Bayerische Akademie der Wissenschaft. Abhandlungen. 1968;**133**:1-86

[28] Siebel W, Höhndorf A, Wendt I. Origin of late Variscan granitoids from NE Bavaria, Germany, exemplified by REE and Nd isotope systematics. Chemical Geology. 1995;**125**:249-270. DOI: 10.1016/0009-2541(95)00083-X

[29] Fiala F. Granitoids of the Slavkovský (Císařský) les Mountains. Sborník Geologických Věd, Geologie. 1968;**14**:93-160

[30] Siebel W, Breiter K, Wendt I, Höhndorf A, Henjes-Kunst F, René M. Petrogenesis of contrasting granitoid plutons in western Bohemia (Czech Republic). Mineralogy and Petrology. 1999;**65**:207-235. DOI: 10.1007/BF01161961

[31] Najman K, Fuksa J, Fulková J, Höschl V, Hron M, Novák K. Krásno – Horní Slavkov, row material Sn-W ore, mineral exploration up to date 30. 09. 1990, Geoindustria, Prague. In: Unpublished Report. 1990. pp. 1-205 (In Czech)

[32] Förster HJ, Tischendorf G, Trumbull RB, Gottesmann B. Late-collisional granite magmatism in the Variscan Erzgebirge, Germany. Journal of Petrolology. 1999;**40**:1613-1645. DOI: 10.1093/petroj/40.11.1579

[33] Pouchou JL, Pichoir F. PAP (φ-ρ-Z) procedure for improved quantitative microanalysis. In: Armstrong JT, editor. Microbeam Analysis. San Francisco: San Francisco Press; 1985. pp. 104-106

[34] Siebel W. Der Leuchtenberger Granit und seine assoziirten magmatischen Gesteine: Zeitliche und stoffliche Entwicklungprozesse im Verlzauf der Entstehung des Nordoberpfalz-Plutons [thesis]. Heidelberg: Ruprecht-Karls Universität; 1993

[35] Boynton WV. Cosmochemistry of the rare earth elements: Meteoritic studies. In: Henderson P, editor. Rare Earth Elements Geochemistry. Amsterdam: Elsevier; 1984. pp. 63-114

[36] Förster HJ, Hunger HJ, Grimm L. Elektronenstrahlmikroanalytische Untersuchungen von Erzmineralen aus Zinn-Lagerstätte Altenberg (Erzgebirge, DDR). 2. Mitteilung: Erster Nachweis der Seltenen Erden Minerale Fluocerit und Bastnäsit. Chemie der Erde. 1987;**46**:73-79

[37] Förster HJ. The chemical composition of uraninite in Variscan granites of the Erzgebirge, Germany. Mineralogical Magazine. 1999;**63**:239-252. DOI: 10.1180/002646199548466

[38] Thomas R, Webster JD. Strong tin enrichment un a pegmatite-forming melt. Mineralium Deposita. 2000;**35**:570-582. DOI: 10.1007/s001260050262

[39] Dolníček Z, René M, Prochaska W, Kovář M. Fluid evolution of the Hub stock, Horní Slavkov-Krásno Sn-W ore district, Bohemian Massif, Czech Republic. Mineralium Deposita. 2012;**47**:821-833. DOI: 10.1007/s00126-012-0400-0

[40] Fleischer M. Relative proportions of the lanthanide in minerals of the bastnaesite group. Canadian Mineralogist. 1978;**16**:361-363

Chapter 4

Deciphering Magmatic Evolution through Zoned Magmatic Enclaves and Composite Dikes: An Example from the Late Cretaceous Taejongdae Granite in Busan, Korea

Mohammed S.M. Adam, Francois Hategekimana, YoungJae Kim and Young-Seog Kim

Abstract

Late Cretaceous granitic intrusions are common in the southeastern Korean Peninsula. Most of these intrusions enclose abundant microgranular enclaves (MEs) and dikes of almost identical age to their plutons. The granitic intrusion in the Taejongdae area encloses a distinct type of enclave known as zoned MEs. The zoned MEs in this region are composed of multiple zones originated from different magmas that have the same origin and age. Several petrological, mineralogical, geochemical, SHRIMP U-Ph age dating, and Lu-Hf isotopic studies have been conducted for the Taejongdae granitoid to identify how different magmas have interacted and formed the zoned MEs. In this chapter, we reviewed previous studies and added some new data to give a comprehensive picture of the Taejongdae granite and emphasize the importance of zoned enclaves and composite dikes in determining the genesis and evolution of granitoids. We interpret that the MEs distributed in the southeastern part of the Korean Peninsula with the age of 75–70 Ma might be closely related to the breakdown of the subducted Izanagi oceanic slab under the Eurasian plate. This tectonic process enhanced the input of new primitive magma into granitic magma chambers and, therefore, restricted the mixing or mingling process, forming the zoned MEs.

Keywords: magmatic enclaves, composite dikes, Gyeongsang Basin, magmatic arc, Cretaceous granite, fractional crystallization

1. Introduction

The interaction of different magmas profoundly influences the genesis and evolution of various granitoids [1–3]. Magmatic enclaves (MEs) within granitic rocks are commonly regarded as a reliable evidence for magma interaction [4]. Consequently, understanding the controlling processes of the development of MEs can thus provide a vital insight into the characteristics of the parental magmas as well as the magmatic processes that drive the evolution of granitic plutons [5–7].

During the Mesozoic, the Korean Peninsula was subjected to three episodes of magmatism: Songnim (Triassic), Daebo (Jurassic), and Bulguksa (Late Cretaceous) [8, 9]. Due to the shallow subduction of the Izanagi plate beneath the northeast Asian continental margin, tectonic and magmatic activities in the Korean Peninsula exhibited a landward-younging trend and extended 1000 km into the continent from the ancient trench during the Early to Middle Jurassic; however, Cretaceous magmatism in the Korean Peninsula exhibited a trenchward-younging trend [10]. The major deformations generated by the subduction of the Izanagi plate are represented by the Tanu-Lu fault in China (to the west) and the Median Tectonic line in Japan (to the east) (**Figure 1a**) [12]. The Korean Peninsula, located between those two main deformation belts, was subjected to left-lateral strike-slip movement. As a result, the Gyeongsan Basin formed in the southeastern part of the Korean Peninsula, along with several minor basins in the southwest (**Figure 1b**). Almost two-thirds of the Cretaceous granitoids are concentrated around the Gyeongsang Basin [10]. Using zircon Hf isotopes [13], the Cretaceous plutonic rocks in the Korean Peninsula were described as Cretaceous–Paleogene granitoids in a magmatic arc (Gyeongsang Arc). They concluded that the Gyeongsang granitoids originated from crustal reworking.

The Late Cretaceous Bulguksa granitoids mainly intruded into the Gyeongsang Basin and recorded variable magmatic fractionation, mixing, and mingling processes. From the previous studies, based on the magmatic events, the Late Cretaceous intrusions in the Gyeongsang Basin could be divided into three main groups: Group 1 consists of granodiorite, enclave-rich porphyritic granite, enclave-poor porphyritic granite, and quartz-monzodiorite, which results from the mixing and mingling of two magmas of different physical properties. Group 2 includes equigranular granite, coarse-grained porphyritic granite, and fine-grained micrographic granite, which result from magma fractionation [14]. Group 3 contains adakite-like granitoids generated by amphibole-dominated fractional crystallization of Bulguksa Arc magma, which is reported in the Jindong area [15]. Most of the Late Cretaceous granitic intrusions in the Gyeongsang Basin enclose microgranular enclaves (MEs) that have almost the same ages as their host granitoids [14].

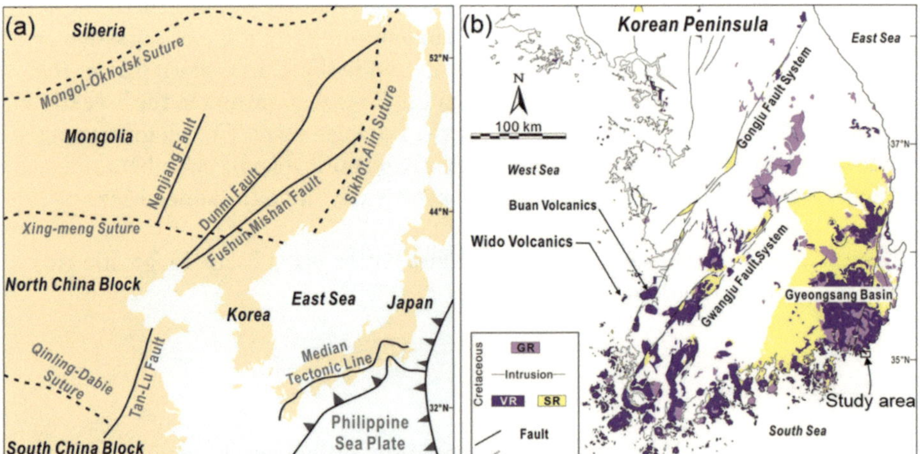

Figure 1.
Tectonic settings of (a) East Asia, and (b) the Korean Peninsula [11]. GR, Cretaceous granitic rocks; VR, Cretaceous volcanic rocks; SR, Cretaceous sedimentary rocks.

The Taejongdae granite, developed in the southeastern part of the Korean Peninsula, contains a distinctive pattern of magmatic enclaves (zoned MEs) made up of multiple different rock zones [16]. The zoned MEs in the Taejongdae granitoid may indicate the magmatic processes that occurred before the intrusion of the granitic pluton in the study area [17]. The first comprehensive petrographic and geochemical study for the Taejondae granitoid and its unique zoned MEs were conducted [16]. In addition, they examined amphibole chemistry within the zoned MEs and the host granite to understand the dynamic process and temperature change during the magmatic evolution. It was concluded that the zoned MEs in the Taejongdae region were formed by mingling and mixing of two magmas: a dioritic magma with a relatively deep crystallization level (7.1–7.7 km) represented by the dioritic zone in the zoned MEs and a shallow-level granitic magma (1.7–2.4 km) represented by the host granite.

Additional data of whole-rock geochemistry, SHRIMP U-Pb zircon age, and Lu-Hf isotope data from zircon grains in the host granite and zoned MEs provide further information on the origin and geochemical characteristics of the dioritic and host granitic magmas and their mutual interaction to produce the zoned MEs [18]. In addition, they compared their geochemical and chronological results with those of other Mesozoic granitoids in the region and proposed some models for the evolution of the Taejongdae granitoids. Furthermore, an ideal composite dike, composed of a felsic granitic interior and mafic margins, which are located approximately 1.2 km from the Taejongdae granitoid to the southeast and hosted in volcano-sedimentary rocks, were reported [19]. The U-Ph zircon age of the felsic granitic interior of the composite dike is almost similar to the age of the Taejongdae granitoid.

In this chapter, we will review the previous studies in the study area to suggest a comprehensive magmatic evolution model and to emphasize the importance of zoned enclaves and composite dikes in determining the magmatic evolution and granitoids' genesis.

2. Petrography and field relationships

Apart from the Taejongdae granite and its associated enclaves in the study area, a variety of mafic dikes are also present, including an ideal composite dike. The enclaves in the study area are divided into two types: simple-type (composed of a single rock type) and composite-type (zoned MEs) (**Figure 2**) [16]. The host granite shows various micro-structural and compositional relationships with the MEs.

The host granite shows a porphyritic texture, miarolitic cavities, and schlieren in certain areas and is composed of albite, orthoclase, and minor amphibole phenocrysts in felsic groundmass minerals. The zoned MEs are characterized by crenulated edges, indicating a mingling relationship between host granite and MEs (**Figure 2a** and **3**) [16, 20]. The simple enclaves, on the other hand, exhibit sharp contact with the host granite, confirming their xenolith origin (**Figure 2b-d**).

The zoned MEs have circular to elliptical shapes and are bordered by three distinct rock zones arranged as follows, from center to rim (**Figure 2a** and **3**):

Zone a is made up of mafic rocks with a porphyritic texture and large amphibole crystals that have been altered into chlorite, as phenocrysts, groundmass minerals, and mesostasis. The groundmass involves plagioclase and anhedral quartz grains. *Zone b* is a mafic rock with fine-to-medium-grained textures with an almost identical mineral composition to zone a. However, the texture of the grains in zone b is slightly finer, and anhedral quartz crystals are more prevalent compared to zone a.

Figure 2.
Various types of MEs hosted in the Taejongdae granite. (a) Zoned ME. (b) Simple ME of type a. (c) Simple ME of type b. (d) Simple ME of type c [16].

Figure 3.
Polished sample of zoned ME showing the relationship between the different zones and the host granite [16].

Zone c is a felsic rock with a coarse grained texture and made up of hornblende, albite, and orthoclase macrocrysts as well as quartz and feldspar microcrysts and opaque minerals. We can see from the polished sample of a zoned ME (**Figure 3**) [16] that the texture of *zone a* changes gradually from being porphyritic to fine in *zone b* and is surrounded by dioritic *zone c*. This gradual change in texture is caused mostly by the rapid cooling

Simple type MEs		
a	**b**	**c**
Hb + Pl + Q	Hb + Pl + Q	Hb + Pl + Q + Kf
Zoned type MEs		
Zone a	*Zone b*	*Zone c*
Hb + Pl + Q + Chl	Hb + Pl + Q	Hb + Pl + Q + Kf
Composite dike		
Mafic margin a	Felsic core	Mafic margin b
Pl + Bi + Q	Q + Pl ± Or	Pl + Bi + Q

Hb, hornblende; Kf, K-feldspar; Mgt, magnetite; Mus, muscovite; Pl, plagioclase; Q, quartz; Chl, chlorite; Bi, biotite.

Table 1.
Summary of the minerals composition of different MEs in the study area (modified from 16).

Figure 4.
Composite dike in the study area. C-D-M1 & C-D-M2 = mafic margins of the composite dike; C-D-F = felsic core of the composite dike [19].

(quenching) of a mafic magma by a felsic dioritic magma. Furthermore, *zone b* contains some feldspar megacrysts from the surrounding dioritic *zone c*, indicating felsic and mafic magma mixed/mingled during the development of the zoned MEs [21]. However, the sharp contact between *zone c* and the host granite indicates that there was no major mixing event between *zone c* and the host granite.

Simple MEs, on the other hand, have fairly angular forms and are relatively small in size compared to zoned MEs (**Figure 2b–d**). The simple enclaves are divided into three types [16] based on the mineral composition and texture (here, denoted by type a, b, and c). Most of the simple MEs (type a, b, and c in hand specimens and under the microscope) exhibit a high similarity with the *Zone a, b,* and *c* of the zonal MEs (**Table 1**). Therefore, it was concluded that most of the simple MEs have resulted from zoned MEs breaking mainly before their full solidification [16].

The composite dike in the study area is emplaced into volcanogenic sedimentary strata and characterized by a felsic interior and mafic margins (**Figure 4**). An abrupt change in chemical composition delineates the magmatic transition between the composite dike core and margins. The mafic margins show an andesitic composition of plagioclase phenocryst in a mafic groundmass, whereas the felsic core has a granitic composition of quartz phenocrysts in a felsic groundmass of plagioclase and quartz in addition to opaque minerals. Petrographically and, to some extent, chemically, the felsic core of the composite dike shows high similarity with the host granite of the study area [19].

3. Methodology

Comprehensive geological mapping was performed in the Taejondae area, including the Taejondae granite (Gamji beach) and the area of the nearby dikes. In addition, SHRIMP U-Pb zircon age, whole-rock geochemistry, and amphibole chemical composition studies were performed.

SHRIMP zircon U-Th-Pb dating was performed on three samples of the host granite, *Zone b* and *c* of the zoned MEs, and the felsic core of the composite dike. The calculated ages represent the zircon crystallization time of the host granitic magma, dioritic magma, mafic enclaves, and the emplacement of the felsic dike, respectively. To extract the zircon grains, samples were crushed for 10 seconds, then sieved, and pulverized for another 10 seconds [18]. They repeated this cycle until enough powder was obtained to collect zircon without breaking the zircon grains. Following that, we collected the zircon grains by applying density-based and magnetic procedures at Pukyong National University. Next, we scanned the zircon grains using cathodoluminescence (CL) to select suitable places for investigation while avoiding alteration and inclusion zones. The zircon grains were put on an epoxy mount, which was subsequently cleaned with petroleum ether and then gold-coated to improve surface conductivity. Following the analytical procedures [22], the SHRIMP method was used to analyze U-Th-Pb from the zircon at the Korean Basic Science Institute (KBSI).

For whole-rock major and trace elements analysis, 24 representative samples (12 for major elements and 12 major and trace elements) were collected from the zoned MEs and host granite (three samples from every zone and the other three from the host granite in different locations) [16, 18]. These samples were trimmed to eliminate weathered surfaces to avoid contamination. At Activation Laboratories Ltd. (ACT LABS) in Canada, the samples were ground to a size of 2 mm, splitted to make the samples representative, and crushed them to sizes >105 μm using a mild steel crusher. Between every two samples, cleaner sand was used. The samples were fused in an induction furnace after being combined with a flux of lithium metaborate and tetraborate. The fused samples were then diluted in a 5% nitric acid solution, thoroughly mixed, and analyzed in ACT LABS in Canada using a Perkin Elmer Sciex ELAN 6000, 6100, or 9000 Inductively Coupled Plasma Mass Spectrometry (ICP-MS). ICP-MS was used to analyze the major and trace elements, including transition metals (Ni, Co, and Cr) and rare earth elements (REEs).

The chemical compositions of amphiboles in a zoned ME and the host granite were investigated [16] at Pusan National University using electron microprobe analysis and CAMECA (France) SX100, 2003. The beam had a diameter of 5 μm, 15 keV accelerating voltage, and 20 nA beam current during analysis. Geothermobarometry of amphiboles is used to determine the P-T condition, H_2O melt, and fO_2 of amphiboles during the crystallization in the zoned MEs and the host granite.

4. Geochemical characteristics and age dating

4.1 Mineral chemistry

The chemical composition of amphibole is susceptible to pressure, temperature, oxygen, and water content variations. Such variations in these characteristics could provide useful information for the evolution of magmatic chamber [23]. One of the most widely used geobarometers is the aluminum content of amphibole. The amphiboles in the *zones a, b,* and *c* of a zoned MEs and the host granite were investigated for major elements using the EPMA technique to comprehend the dynamic operation and the change in temperature during the formation of the host granite in the Taejongdae region [16, 24]. The P–T conditions, H_2O melt, and fO_2 of the amphibole in zones (b) and (c), as well as the host granite, are calculated using the proposed formula [25]. The geobarometer calculations revealed that zone (b) amphibole crystallized at T = 798–827°C, P = 134–174 MPa, NNO = 0.9–0.5, fO_2 = 13.–12.8, H_2O melt = 6.1–6.9 wt%, which corresponds to the continental depth = 5.1–6.6 km, whereas zone (c) amphibole crystallized at T = 877–900°C, P = 187–205 MPa, ΔNNO: 0.4–0.7, fO_2: -11.6 – -11.2, H_2Omelt: 4.6–5.3 wt% and continental depth of 7.1–7.7 km. The host granite exhibits the following crystallization state; T = 727–775°C, P = 45–65 MPa, NNO = 1.9–2.2, fO_2 = 13.4–12.5, H_2O melt = 4.4–4.5 wt%, and the continental depth = 1.7–2.4 km. From these results, we concluded that the zoned MEs in the Taejongdae area were made by mixing and mingling of two different magmas [16]: the first is a dioritic magma with a relatively deep crystallization level (7.1–7.7 km), represented by the dioritic zone c in the zoned MEs, and the second is a shallow level granitic magma (1.7–2.4 km). We also reported the geochemical characteristics and origins of the two magmas (dioritic and host granite magmas) and how they interacted to produce the zoned MEs [18]. Furthermore, we compared their geochemical and temporal results with other Mesozoic granitoids in the research area and provided proper models to demonstrate the evolution mechanism of the Taejongdae granitoid.

4.2 Age dating and Lu-Hf isotope analysis

The host granite and zoned MEs were studied using SHRIMP U-Pb zircon age, and Lu-Hf isotope analysis [18]. In addition, SHRIMP U-Pb zircon age dating from the felsic core of the composite dike was conducted [19]. The age dating results indicate that all the rock samples from the host granite, zone b and c of the zoned MEs and the core of the composite dikes displayed $^{206}Pb/^{238}U$ age of 72.3 ± 0.7 Ma, 71.5 ± 0.7 Ma,

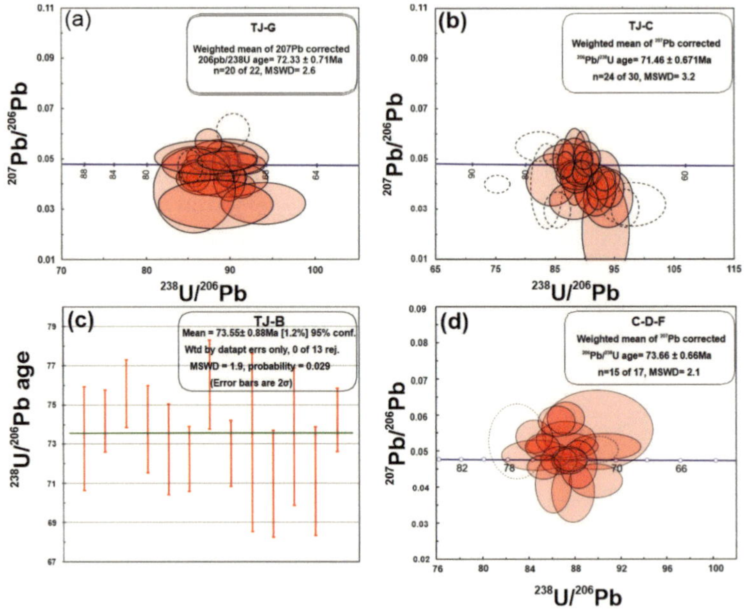

Figure 5.
Concordia diagrams of zircon U-Pb analytical results for samples from the host granite, zoned MEs [18] and core of composite dike [19]. (a) Host granite. (b) Zone c. (c) Recalculated concordia age of the zone b zircon. (d) Felsic core of the composite dike [19].

73.5 ± 0.9 Ma, and 73.66 ± 0.66 Ma, respectively (**Figure 5**). Whereas the Lu-Hf isotope data from the host granite and zoned MEs show nearly identical Hf(t) (Hf at the time zircon crystallized) values ranging from −2.0 to +17.4 but grouped around +5, while TDM (depleted mantle model ages) are clustered around 600 Ma.

4.3 Whole-rock geochemistry

The granitic magma of the host granite and the dioritic magma, as seen in zone c of the zoned MEs, were likely combined to form the zoned MEs in the Taejongdae area [16]. The chemical properties of representative samples from the host granite and zone c of the zoned MEs have been studied [18] to determine the characteristics of those two magmas.

From using TAS (SiO_2 vs. Na_2O + K_2O; [26]), AFM diagram [27], K_2O versus SiO2, and A/CN-A/NK plots, the host granite samples are identified as high-K calc-alkaline, peraluminous granites (**Figure 6**). Furthermore, the primitive mantle and chondrite normalized REE and trace elements of the host granite exhibit negative Eu anomalies and depleted Nb, P, and Ti patterns (**Figure 7**). Such geochemical characteristics of the granite in the Gyeongsang Basin are typical of the Bulguksa granite. Furthermore, in terms of chemical characteristics and crystallization ages, the Taejongdae host granite is similar to Group IV Cretaceous granitoids (according to the classification of [10]) (**Figure 7a** and **8k**).

Samples from zone c of the zoned MEs (representative of the dioritic magma) exhibit similar characteristics to the Bulguksa granite of dioritic, calc-alkaline, peraluminous magma (**Figures 6** and **7a–c**). The major element chemistry of zone c exhibits some adakitic signatures with SiO2 > 56%, Na2O > 3.5%, Al2O3 > 15%, K2O/Na2O ratios ~0.4, and a positive-to-flat Eu anomaly [31]. Adakite and Archaean tonalite–trondhjemite–granodiorite (TTG) are generally known to originate from slab-melting [32]. However, the trace element chemistry of zone c samples show low Sr./Y and La/Tb

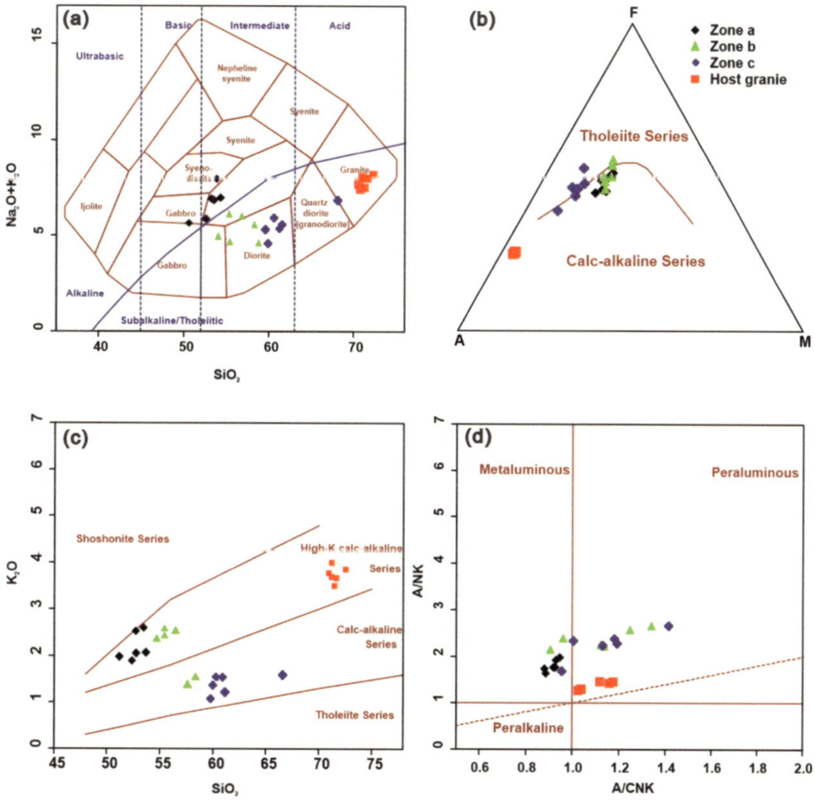

Figure 6.
Geochemical classification of zoned MEs and the host granite samples from the study area [16, 18]. (a) Classification of the zones of zoned MEs and the host granite rocks based on the total alkali versus silica of TAS diagram (SiO2 versus Na2O + K2O; [25]. (b) AFM diagram showing the boundary between tholeiitic and calc-alkaline fields; the majority of the samples plotted in the calc-alcaline field [27]. (c) Classification of the samples using SiO2 versus K2O diagram, with fields defined by [28]. d) Plot of a/NK versus a/CNK for the samples. Zones b, c, and the host granite are weak peraluminous, whereas zone a is metaluminous. A/NK – molar ratio of Al2O3/(Na2O + K2O), a/CNK = molar ratio of Al2O3/(CaO + Na2O + K2O).

ratios, which is not consistent with the general feature of adakites and TTG. Thus, it is unlikely that rocks in zone c are adakites. Therefore, we classified zone c dioritic magma as originating from Bulguksa granitic magma [18]. However, to distinguish from the host granite magma, we described zone c dioritic magma as low-K magma.

5. Evolution of the Taejongdae granite

5.1 Origin of the Taejongdae granite

The Bulguksa granitoids in the Gyeongsang Basin (Eonyang area) are classified into several groups and described as porphyritic granitic plutons, or "enclave-rich porphyritic granite (ERPG)" [14]. It is believed that the ERPG was formed by the mixing and mingling of magmas generated from melting of the base of the continental crust (igneous origin) by basaltic magma, which in turn was produced from the upwelling of the asthenosphere beneath the eastern part of the Eurasian continent. The upwelling of the asthenosphere was triggered by the formation of a slab window in the subducted Izanagi–Pacific plate beneath the eastern part of the Asian continent [14, 33].

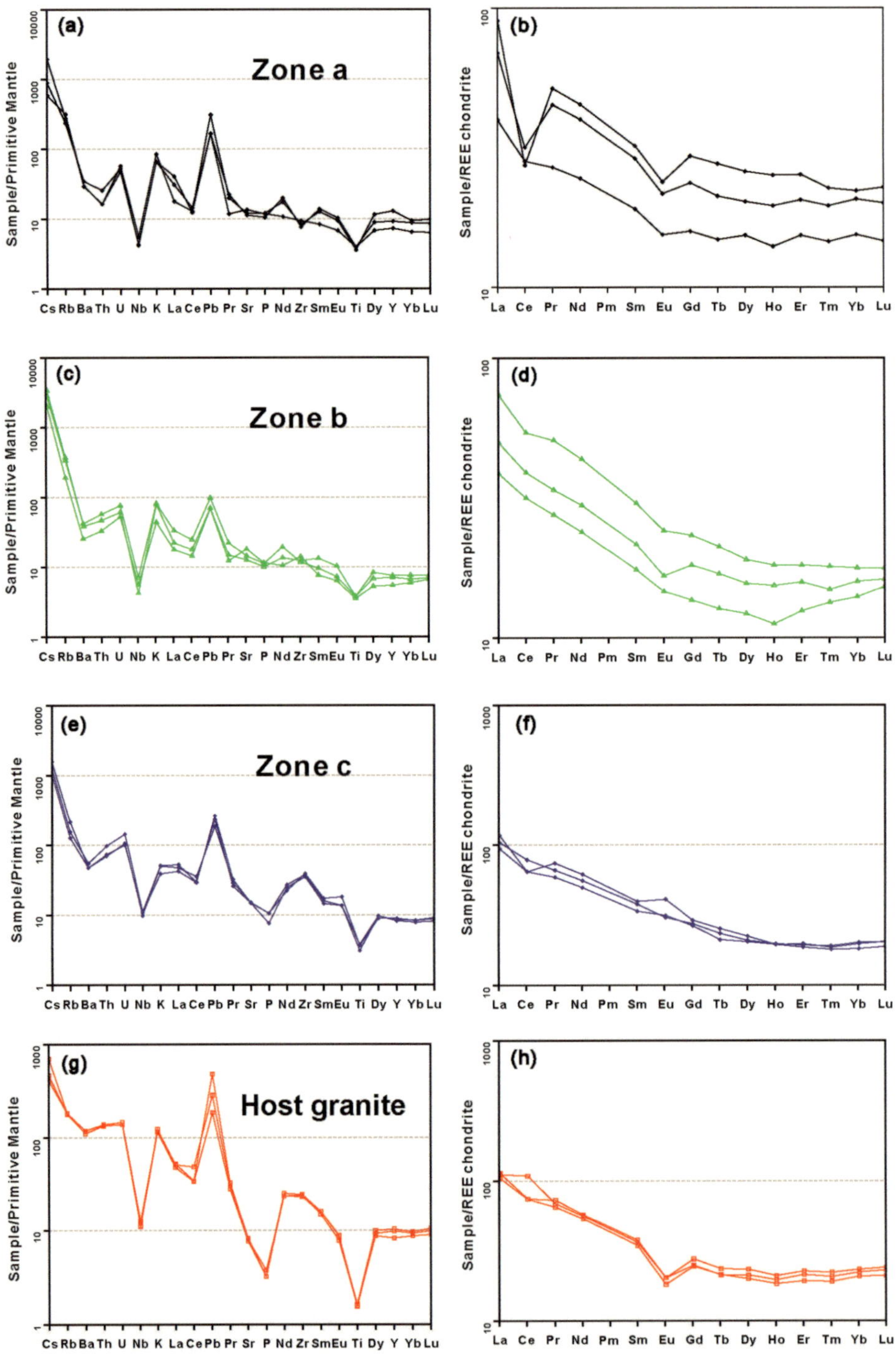

Figure 7.

Whole-rock REE and trace-elements concentrations of the zones a, b, and c and the host granite, normalized to the primitive mantle [29] and chondrite normalized values [30]. (a–b) Zone a samples. (c–d) Zone b samples. (e–f) Zone c samples show positive-to-no anomaly. (g–h) Host granite samples. Data from [18].

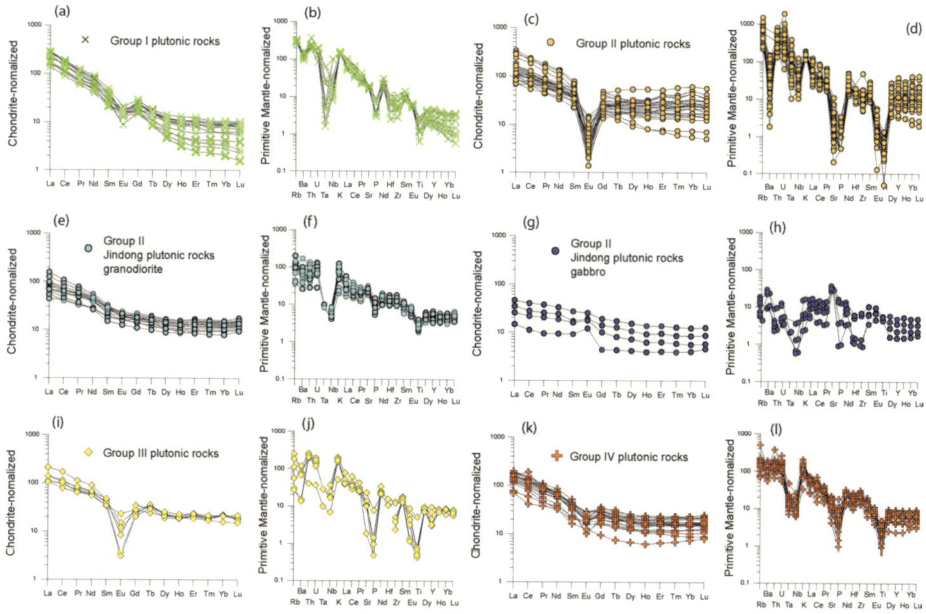

Figure 8.
Chondrite-normalized rare earth elements and primitive mantle-normalized trace elements of the cretaceous plutons in the Korean Peninsula [10].

The results of age dating and Lu-Hf isotopes for the Taejongdae granitoid also indicate Late Cretaceous age and the same mixed origin with $^{176}Hf/^{177}Hf$ ratios of 0.282670–0.282978 and εHf (t) values ranging from −2 to 8.8 [18]. These data implied that the host granite was formed from a combination of primitive basaltic melts and preexisting crustal rocks of the Paleozoic to Neoproterozoic age [18]. Therefore, the host granite in the Taejongdae area could be classified as Bulguksa granitoid and further classified as the ERPG.

Among the four classified groups of Cretaceous granitoids [10], the dioritic magma of zone c (low-K magma) has similar chemical characteristics to group II granitoids and gabbro (Jindong plutonic rocks), despite their age difference. Both Taejongdae and Jindong granitoids are calc-alkaline and are slightly enriched in LREEs; they also exhibit flat-to-positive Eu anomalies and depleted patterns in P, Nb, and Ti. Such calc-alkaline-like magma with adakitic signatures has also been reported in two granitic intrusions in the southern part of the Korean Peninsula: (i) the Cretaceous Jindong granite in the Gyeongsang Basin and (ii) the Bongnae granitic intrusion in the southeastern part of the Korean Peninsula. From a detailed study of the Jindong and Bongnae granitoids [15], it is concluded that the Cretaceous Jindong pluton was formed neither from adakitic nor TTG magma but from amphibole-dominated fractional crystallization of hydrous Bulguksa-like arc magma. In contrast, the Triassic Bongnae plutons formed from a K-rich C-type adakite-like magma.

The $^{176}Hf/^{177}Hf$ ratios of 0.282801–0.282977 and εHf (t) values of 2.6–8.8 of the low-K magma [18] confirm that they have the same source as the host granite. However, different fractional crystallization processes of the same magma could generate different products. Therefore, it was suggested that the low-K magma represents arc magma generated in the Late Cretaceous as the host granite magma (calc-alkaline hydrous Bulguksa-like arc magma) [18]; however, it was crystallized by amphibole-dominated fractional crystallization as Jindong granitoid.

5.2 Evolution of the Taejongdae granite and formation of the zoned MEs

We suggested a model of two stages for the evolution of the Taejongdae granite and the formation of the zoned MEs, by an interaction between the two magmas (the host granite and low-K magma) [16]. In addition, $^{206}Pb/^{238}U$ age, whole-rock geochemical data, and the Lu-Hf isotope analysis from zircon support the two-stage model and a new possibility that the host granite magma may have formed due to rhyolitic melt segregation from the low-K magma [18]. However, rhyolitic melt segregation requires regional compaction [34]. Furthermore, from the study of the composite dike [19], the felsic core of the composite dike has the same emplacement age as the host granite, and the felsic core of the composite dike formation requires an extensional setting. Therefore, the rhyolitic melt segregation model has no enough solid evidences be considered as the formation mechanism for the Taejongdae granite.

5.2.1 Two-stage model

In the first stage, amphibole-dominated fractional crystallization of calc-alkaline arc magma generated a low-K magma at 7.1–7.7 km depth [16]. At 73.55 ± 088 Ma, a trachy-andesitic magma was injected into the low-K granodioritic magma and formed magmatic enclaves. Due to the temperature difference between the trachy-andesitic and the low-K magma, zone b cooled rapidly while zone a cooled gradually. The tecture and

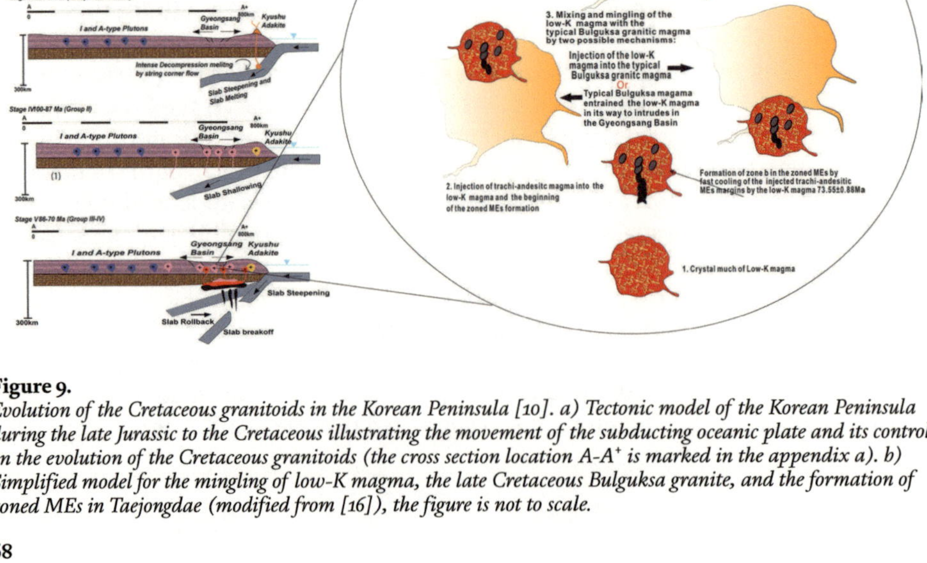

Figure 9.
Evolution of the Cretaceous granitoids in the Korean Peninsula [10]. a) Tectonic model of the Korean Peninsula during the late Jurassic to the Cretaceous illustrating the movement of the subducting oceanic plate and its control on the evolution of the Cretaceous granitoids (the cross section location A-A^{+} *is marked in the appendix a). b) Simplified model for the mingling of low-K magma, the late Cretaceous Bulguksa granite, and the formation of zoned MEs in Taejongdae (modified from [16]), the figure is not to scale.*

zircon age of zone b reflect the trachy-andesitic magma injection sequence. At the same time or slightly later, another Bulguksa calk-alkaline arc magma developed through fractional crystallization and reached the surface (which represents the host granite).

In the second stage, the two magmas could be combined through two possible manners. The first possibility is that the typical Bulguksa magma of the host granite intruded the basin and entrained the low-K magma and its magmatic enclaves as enclaves (second round of enclaves formation) without mixing (due to the physical properties differences) and solidified at a shallow level in the Gyeongsang Basin as the Taejeongdae granite. The second possibility is that the low-K magma with its trachy-andesitic enclaves were ascended and injected into the host granite magma, and consolidated at a shallow level in the Gyeongsang Basin (72.33 ± 0.71 Ma) (**Figure 9c**).

6. Tectonomagmatic setting

The diorite-granodiorite-granite plutons of the southeastern (or southern) part of the Korean Peninsula, including the Taejongdae region, are calc-alkalic granitoids with or without MEs. By mixing/mingling magmas, more mafic types (Group I of [14]) were generated. In contrast, granodiorite-granite plutons (Group II of [14]) were formed by the fractional crystallization of a parent magma. According to their geochemical characteristics, the 75–70 Ma diorite-granite plutons are high-K and calc-alkaline rocks related to subduction [10, 12].

The tectonic and magmatic activities, in the Korean Peninsula, during the Cretaceous (120–70 Ma) show an oceanward-younging trend and spread out approximately ~800 km into the continent from the subduction zone (the ancient trench) (**Figure 9a**). The magma from the old trench may indicate shallow subduction of the Izanagi plate [10, 35]. In the southeastern part of the Korean Peninsula, the tectonic and magmatic activities during 75–70 Ma might be closely related to slab steeping due to slab rollback of the Izanagi plate, leading to lithospheric and/or crustal thinning and oceanward arc migration [10]. The subduction slab rollback represents one of the main factors controlling the stress state transition from compression to extension [36]. During the same duration, the breakdown of the subducted oceanic slab created a slab window, having a narrow gap and permitting the asthenospheric upwelling (**Figure 9a**) [14, 37].

Therefore, based on our U-Pb zircon age and Lu-Hf isotope data, we suggest that these tectonomagmatic processes resulted in the emplacement of the Late Cretaceous (75–70 Ma) mafic and MEs-bearing granitoid plutons, having crustal isotopic signatures. The abundant MEs and dikes distribution in the southeastern part of the Korean Peninsula, during 75–70 Ma (appendix A), might be closely related to the breakdown of the subducted Izanagi oceanic slab under the Eurasian plate.

7. Conclusions

Taejongdae area, located in the southeastern part of the Korean Peninsula, was selected in this chapter as a case study for deciphering the magmatic evolution through the analysis of zoned magmatic enclaves and composite dikes.

The Taejongdae granite is believed to have originated from the mixing and mingling of magmas generated from the melting of the base of the continental crust by basaltic magma, which was produced from the upwelling of the asthenosphere beneath the eastern part of the Eurasian continent. The host granite was formed from a combination of primitive basaltic melts and preexisting crustal rocks of Paleozoic to Neoproterozoic age. The Taejongdae granite is classified as Bulguksa granitoid and further classified as enclave-rich porphyritic granite (ERPG).

The zoned MEs were formed in two stages due to the interaction of two magmas, the host granite and low-K magma. The low-K magma represents arc magma generated in the Late Cretaceous (similar to the host granite magma); however, it was crystallized by amphibole-dominated fractional crystallization.

The intense concentration of the Cretaceous MEs and mafic dikes in the southernmost part of the Korean Peninsula was primarily controlled by magma mixing induced by asthenospheric upwelling during the subduction of the Izanagi oceanic plate under the eastern margin of the Eurasian plate. The zoned MEs and composite dikes in the Taejongdae area indicates the characteristic magmatic process that took place in subduction zone setting. This study indicates that detailed analyses of enclaves and dikes could give very useful information on the understanding of the tectonic evolution as well as the magmatic interaction of the region.

Acknowledgements

This research was financed by the BK21 plus Project of the Graduate School of Earth Environmental Sciences System (21A20151713014) and a grant from the Korea Institute of Energy Technology Evaluation and Planning (KETEP) awarded by the Korea government (MOTIE) (20201510100020). We wish to express our appreciation to the sponsors and GSGR members for their assistance and support.

Conflict of interest

The authors declare no conflict of interest.

A. Appendix

Ages of the plutons from the previous studies. The Late Cretaceous granitoids (with red colors) are concentrated in the southern part of the Korean Peninsula (Modified from [10]).

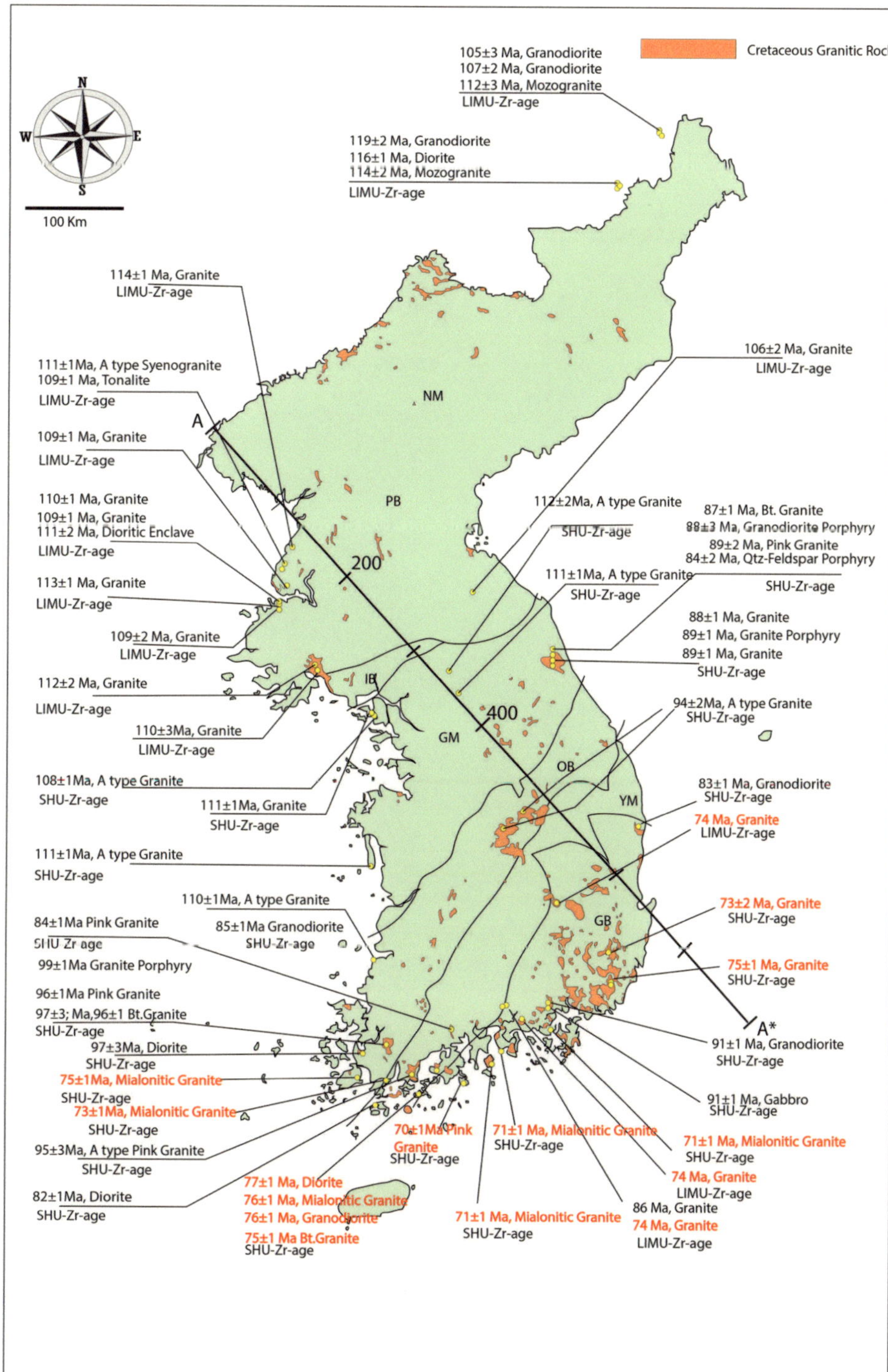
Cretaceous Granitic Rock
N
W
E
S
100 Km
105±3 Ma, Granodiorite
107±2 Ma, Granodiorite
112±3 Ma, Mozogranite
LIMU-Zr-age
119±2 Ma, Granodiorite
116±1 Ma, Diorite
114±2 Ma, Mozogranite
LIMU-Zr-age
114±1 Ma, Granite
LIMU-Zr-age
111±1Ma, A type Syenogranite
109±1 Ma, Tonalite
LIMU-Zr-age
106±2 Ma, Granite
LIMU-Zr-age
NM
A
109±1 Ma, Granite
LIMU-Zr-age
PB
110±1 Ma, Granite
109±1 Ma, Granite
111±2 Ma, Dioritic Enclave
LIMU-Zr-age
112±2Ma, A type Granite
SHU-Zr-age
87±1 Ma, Bt. Granite
88±3 Ma, Granodiorite Porphyry
89±2 Ma, Pink Granite
84±2 Ma, Qtz-Feldspar Porphyry
SHU-Zr-age
200
113±1 Ma, Granite
LIMU-Zr-age
111±1Ma, A type Granite
SHU-Zr-age
109±2 Ma, Granite
LIMU-Zr-age
88±1 Ma, Granite
89±1 Ma, Granite Porphyry
89±1 Ma, Granite
SHU-Zr-age
112±2 Ma, Granite
LIMU-Zr-age
IB
400
GM
94±2Ma, A type Granite
SHU-Zr-age
110±3Ma, Granite
LIMU-Zr-age
OB
108±1Ma, A type Granite
SHU-Zr-age
111±1Ma, Granite
SHU-Zr-age
YM
83±1 Ma, Granodiorite
SHU-Zr-age
74 Ma, Granite
LIMU-Zr-age
111±1Ma, A type Granite
SHU-Zr-age
110±1Ma, A type Granite
73±2 Ma, Granite
SHU-Zr-age
84±1Ma Pink Granite
SHU Zr age
85±1Ma Granodiorite
SHU-Zr-age
GB
99±1Ma Granite Porphyry
75±1 Ma, Granite
SHU-Zr-age
96±1Ma Pink Granite
97±3; Ma,96±1 Bt.Granite
SHU-Zr-age
A*
97±3Ma, Diorite
SHU-Zr-age
91±1 Ma, Granodiorite
SHU-Zr-age
75±1Ma, Mialonitic Granite
SHU-Zr-age
91±1 Ma, Gabbro
SHU-Zr-age
73±1Ma, Mialonitic Granite
SHU-Zr-age
70±1Ma Pink Granite
SHU-Zr-age
71±1 Ma, Mialonitic Granite
SHU-Zr-age
71±1 Ma, Mialonitic Granite
SHU-Zr-age
95±3Ma, A type Pink Granite
SHU-Zr-age
74 Ma, Granite
LIMU-Zr-age
82±1Ma, Diorite
SHU-Zr-age
77±1 Ma, Diorite
76±1 Ma, Mialonitic Granite
76±1 Ma, Granodiorite
75±1 Ma Bt.Granite
SHU-Zr-age
71±1 Ma, Mialonitic Granite
SHU-Zr-age
86 Ma, Granite
74 Ma, Granite
LIMU-Zr-age

Author details

Mohammed S.M. Adam, Francois Hategekimana, YoungJae Kim
and Young-Seog Kim*
Division of Earth and Environmental System Science, GSGR, Pukyong National University, Busan, Republic of Korea

*Address all correspondence to: ysk7909@pknu.ac.kr

References

[1] Barbarin B. A review of the relationships between granitoid types, their origins and their geodynamic environments. Lithos. 1999;**46**(3):605-626

[2] Barbarin B. Mafic magmatic enclaves and mafic rocks associated with some granitoids of the 653 Central Sierra Nevada batholith, California: Nature, origin, and relations with the hosts. Lithos. 2005;**80**(1-4):155-177

[3] Słaby E, Martin H. Mafic and felsic magma interaction in granites: The Hercynian Karkonosze pluton (Sudetes, bohemian massif). Journal of Petrology. 2008;**49**(2):353-391

[4] Didier J. Contribution of enclave studies to the understanding of origin and evolution of granitic magmas. Geologische Rundschau. 1987;**76**(1):41-50

[5] Didier J, Renouf JT. Granites and their enclaves: The Bearing of Enclaves on the Origin of Granites. Vol. 3 Amsterdam: Elsevier; 1973. p. xiv+393

[6] Didier J, Barbarin B. The different types of enclaves in granites-nomenclature. In: Enclaves and Granite Petrology. Amsterdam: Elsevier; 1991;**13**:19-23

[7] Barbarin B, Didier J. Genesis and evolution of mafic MEs through various types of interaction between coexisting felsic and mafic magmas. Earth and Environmental Science Transactions of the Royal Society of Edinburgh. 1992;**83**(1-2):145-153

[8] Kobayashi T. Geology of South Korea with special reference to the limestone plateau of Kogendo. Journal of Faculty of Science, Imperial University of Tokyo, Section 2. 1953;**8**(4):145-233

[9] Reedman AJ, Um SH. Geology of Korea. Seoul. Korea: Korean Institute of Energy resource; 1975

[10] Kim SW, Kwon S, Park SI, Lee C, Cho DL, Lee HJ, et al. SHRIMP U–Pb dating and geochemistry of the cretaceous plutonic rocks in the Korean peninsula: A new tectonic model of the cretaceous Korean peninsula. Lithos. 2016;**262**:88-106

[11] Ko K, Kim SW, Lee HJ, Hwang IG, Kim BC, Kee WS, et al. Soft sediment deformation structures in a lacustrine sedimentary succession induced by volcano-tectonic activities: An example from the cretaceous Beolgeumri formation, Wido Volcanics, Korea. Sedimentary Geology. 2017;**358**:197-209

[12] Otsuki K. Plate Tectonics of eastern Eurasia in the light of fault systems. Scientific Reports. Japan: Tohoku Univesristy; 1985;**55**:141-251

[13] Cheong ACS, Jo HJ. Crustal evolution in the Gyeongsang arc, Southeastern Korea: Geochronological, geochemical and Sr-Nd-Hf isotopic constraints from granitoid rocks. American Journal of Science. 2017;**317**(3):369-410

[14] Hwang BH. Petrogenesis of the Eonyang granitoids, SE Korea: New SHRIMP-RG zircon U-Pb age and whole-rock geochemical data. International Geology Review. 2012;**54**(1):51-66

[15] Oh JI, Choi SH, Yi K. Origin of adakite-like plutons in southern Korea. Lithos. 2016;**262**:620-635

[16] Adam MS, Kim T, Song YS, Kim YS. Occurrence and origin of the zoned

MEs within the cretaceous granite in Taejongdae, SE Korea. Lithos. 2019;**324**:537-550

[17] Morgavi D, Laumonier M, Petrelli M, Dingwell DB. Decrypting magma mixing in igneous systems. Reviews in Mineralogy and Geochemistry. 2022;**87**(1):607-638

[18] Adam MS, Kim SW, KimT, Naik SP, Cho K, Kim YS. Constraining Mixing and Mingling processes from Zoned Magmatic Enclaves: An example from the Taejongdae granite in Busan, Korea [Under review]. Lithos

[19] Adam MS, Kim T, Kim YS. Formation mechanism of the composite dyke in the Taejongdae area, SE Korea [in process]. In: Earth and Environmental Sciences. Pukyong National University

[20] Vernon RH, Etheridge MA, Wall VJ. Shape and microstructure of microgranitoid enclaves: Indicators of magma mingling and flow. Lithos. 1988;**22**(1):1-11

[21] Griffin WL, Wang X, Jackson SE, Pearson NJ, O'Reilly SY, Xu X, et al. Zircon chemistry and magma mixing, SE China: in-situ analysis of Hf isotopes. Tonglu and Pingtan igneous complexes. Lithos. 2002;**61**(3-4):237-269

[22] Williams IS. U-Th-Pb geochronology by ion microprobe. Reviews in Economic Geology. 1998;**v**:7

[23] Johnson MC, Rutherford MJ. Experimental calibration of the aluminum-in-hornblende geobarometer with application to Long Valley caldera (California) volcanic rocks. Geology. 1989;**17**(9):837-841

[24] Gourgaud A. Comagmatic enclaves in lavas from the Mont-Dore composite volcano, Massif Central, France. In: Enclaves and Granite Petrology. Amsterdam: Elsevier; 1991;**13**:221-233

[25] Ridolfi F, Renzulli A, Puerini M. Stability and chemical equilibrium of amphibole in calc-alkaline magmas: An overview, new thermobarometric formulations and application to subduction-related volcanoes. Contributions to Mineralogy and Petrology. 2010;**160**:45-66

[26] Cox KG, Bell JD, Pankhurst RJ. The Interpretation of Igneous Rocks. London: George Allen & Unwin; 1979

[27] Irvine TN, Baragar WRA. A guide to the chemical classification of the common volcanic rocks. Canadian Journal of Earth Sciences. 1971;**8**(5):523-548

[28] Peccerillo A, Taylor SR. Geochemistry of Eocene calc-alkaline volcanic rocks from the Kastamonu area, northern Turkey. Contributions to Mineralogy and Petrology. 1976;**58**:63-81

[29] Sun SS, McDonough WF. Chemical and isotopic systematics of oceanic basalts: Implications for mantle composition and processes. Geological Society, London, Special Publications. 1989;**42**(1):313-345

[30] Boynton WV. Cosmochemistry of the rare earth elements: Meteorite studies. In: Developments in geochemistry. Amsterdam: Elsevier. 1984;**2**:63-114

[31] Castillo PR. An overview of adakite petrogenesis. Chinese Science Bulletin. 2006;**51**:257-268

[32] Defant MJ, Drummond MS. Derivation of some modern arc magmas by melting of young subducted lithosphere. Nature. 1990;**347**(6294):662-665

[33] JWA, Y. J. Temporal, spatial and geochemical discriminations of granitoids in South Korea. Resource Geology. 1998;**48**(4):273-284

[34] McKenzie D. The extraction of magma from the crust and mantle. Earth and Planetary Science Letters. 1985;**74**(1):81-91

[35] Kee WS, Won Kim S, Jeong YJ, Kwon S. Characteristics of Jurassic continental arc magmatism in South Korea: Tectonic implications. The Journal of Geology. 2010;**118**(3):305-323

[36] Kay SM, Godoy E, Kurtz A. Episodic arc migration, crustal thickening, subduction erosion, and magmatism in the south-Central Andes. Geological Society of America Bulletin. 2005;**117**(1-2):67-88

[37] Suga K, Yeh MW. Secular variation of early cretaceous granitoids in Kyushu, SW Japan: The role of mélange rocks as a possible magma source. Frontiers in Earth Science. 2020;**8**:95

Chapter 5

Examining the Effect of Powder Factor Variability on Granite Productivity

Luqman Kareem Salati and Moses Shola Adeyemo

Abstract

This research seeks to examine the effect of powder factor variability on granite productivity during its quarrying. Schmidt hammer was used for the in-situ determination of rock hardness. Uniaxial compressive strength (UCS) of in situ rock was estimated from the values obtained from Schmidt hammer rebound hardness test and its density determined from laboratory test. After preliminary field studies, ten (10) blasts with varied powder factors were studied and their overall effects on granite productivity examined. Three (3) rock samples were carefully collected from the quarrying site and subjected to laboratory analysis for UCS and bulk density tests. With spacing and burden kept between 1.7 m and 1.8 m and stemming height also varied between 1.5 m and 2 m, charge columns of between 4.5 m to 6.5 m were maintained, while number of holes drilled per blast was between 64 and 88. Results obtained from the test revealed that the average UCS of the granite samples was 80.67 MN/m^2 while the average bulk density was 2465.67 kg/m^3. Therefore, considering ten (10) blasts with varied powder factors of between 0.77 kg/m^3 and 0.97 kg/m^3, total volumes of rock of between 1109.76 m^3 and 2280.96 m^3 was produced. Hence, varied powder factors have been found to have varying effects on rock fragmentation sizes and by extension, granite productivity.

Keywords: powder factor variability, explosives, granite productivity, blasting, fragmentation

1. Introduction

Efficiency of blasting operations in underground and surface mines determines, to a large extent, utilization of equipment, productivity and economics. Proper fragmentation of blasted rocks improves the efficiency of downstream operations by loading and crushing to desired sizes. An optimal blast not only results in proper fragmentation but also reduces undesirable effects in ground vibration, fly rock and formation of toe in quarry benches. Drilling and blasting are the first unit operations in the mining process and have a major impact on the performance and cost of subsequent unit operations [1–3].

According to Salati and Mark [4], powder factor can be defined as the quantity of explosives needed to fragment a unit cubic metre of rock ($1m^3$). Hence, optimum powder factor results in good fragmentation, having less throw and less ground vibration. It can serve as an indicator for rock hardness, cost of explosives used or as a guide to shot firing plan [5].

Improved fragmentation gives loading equipment a higher rate of productivity; hence, it results in lower cost per tonne or cubic yard moved. The effect of wear and tear also decreases giving lower operating cost per hour under similar condition of haul, lift, size and type of truck. Haul/load road condition, truck production per hour also increase with greater degree of fragment due to faster shovel or loader longing rate and a decrease in bridging at the crusher. Therefore, there is a consequent decrease in cycle time. Fragmentation optimization involves breaking of rock to ensure quality control, safe, consistent and efficient blasting [6]. Subsequently, boulder or the opposite, excess fines, result from poorly selected drilling and blasting patterns. A well selected pattern would produce fragmentation that can be accommodated by available loading and hauling equipment and crushing plant with little or no need for secondary blasting. Therefore, it is a well acknowledged fact that the performance of mining operations such as excavation and crushing reeves on fragmentation has been pre-conditioned by blast designs [6–8].

The effectiveness of hard rock blasting is measured with two basics indices namely, oversize generation and blast hole productivity. Cost per tonne of rock blasted is another index that measures the effectiveness of blasting and is dependent on rockiness and blast design parameter such as hole diameter, burden, spaces among others [9, 10]. Such parameters differ from one mine to the other and some of the blast design parameters could be regulated to deliver the desired blasting effectiveness. The individual influence of the determinant parameter on blasting has been studied by several authors but their cumulative influence on the same is yet to be formulated. However, the huge statistical data generated from the well organized and documented large scale hard rock surface mines operating variables condition worldwide constitute the only readily available resources which could be used for the analysis and regression model of indices that determine effectiveness of blasting of rock blasted on uncontrollable and controllable blasting parameters [1]. Efficiency of drilling and blasting operations must contribute to the best overall economics of a quarry [11, 12]; therefore, variability of powder factor has potentials to improve surface mines' productivity [13, 14]. Hence, there is need to study the effect of powder factor variability in the productivity of granite quarrying. The study is an attempt to achieve the following specific objectives:

i. To determine the relationship between the powder factor and uniaxial compressive strength at the quarry.

ii. To determine the effect of varying powder factor in the fragmentation of rock; and

iii. To recommend ways to improve the productivity of granite in the quarry.

The appraised parameters would give optimum blasting results through the regression model generated using indices such as oversize generation and geometric volume of the blasted rock on blast design parameters.

2. Materials, methods used

2.1 Climate, vegetation and relief of typical granite areas

A study carried out by Abaje and Oladipo [15] shows that the temperatures of granite mineralized areas could increase from 0.2 to 0.5°C. A case of rainfall trend and variation characteristics across Kaduna State, North-western Nigeria, using 11 selected stations in the Southern, Central and Northern parts of the State for a period of 50 years (1966–2015) was carried out and the climatic condition of the study revealed that its Southern part, has the highest total rainfall, yet there was no significant trend in the five decadal periods examined. From the early 1970s to the late 1990s the rainfall was below the long-term mean. The rainfall of the remaining years nearly approximates the long-term mean [16]. Local spatial patterning of most granitic areas partly describes their spatial heterogeneity and related to alternation between vegetated and bare areas which are commonly found in savanna systems [17].

Ladan [18] also shows that granitic areas which have natural rain forest vegetation often fall within the Savanna. Their two distinct characteristic seasons are rainy and dry, while their annual rainfall is between 80 cm and 100 cm. Their temperature variation throughout the year could vary from 20° to 27°. The vegetation in such area may be Guinea Savannah and can be characterized by long and short grasses as well as low trees and shrubs. Their sparse distribution is sometimes accompanied by dense vegetation along streams and river channels [19].

2.2 Brief geology granite areas

The areas consist of rocks that range in age from Pre-Cambrian to Lower Paleozoic and Quaternary period. Four groups of rocks can be distinguished for the Basement Complex Terrain in the area. The crystalline basement rocks which consist of gneisses and migmatites with different varieties of the gneisses like the banded gneiss, granite gneiss, biotite gneiss, hornblende gneiss and ortho genesis [20]. **Figure 1** shows the exposure of a typical granite outcrop.

Figure 1.
Exposure of a typical Quarry's granite outcrop showing its relief.

In the older granite underlain by the Precambrian Basement Complex rock, porphyritic granite could be the most common basement rock intruding both magnetite and metasediments. Pegmatites are widely distributed throughout the Precambrian Basement Complex of the area studied by Abere *et al.* [21]. However, the extremely coarse igneous bodies which are closely weathered and spaced to large masses plutonic rock often consist of quartz, feldspars and muscovite [22]. They are elliptical to elongated shape, which is seen to be elevated to the forsian emplacement.

2.3 Field studies and data collection

The operation of the selected granite site was studied in order to obtain information about the various explosives, drilling machines, blasting parameters such as burden, spacing and depth of hole. Their physical condition and quantity available were studied. Observation was also made to estimate the size of fragmented rocks. Three samples of blasted boulders were collected at three different faces of the quarry in the study area. The coordinates of each location were taken and recorded with the aid of global positioning system (GPS).

2.4 Drilling/blasting and powder factor variability procedures

As shown in **Tables 1–10**, the various drilling and blasting parameters used at Tutu quarry including the parameters for burden and spacing, depth of blast holes, column

S/N	Parameters	Numeric value	
1	Explosive used	High Explosive + ANFO	
2	Drilling pattern	Staggered	
3	Hole depth	6.30	M
4	Diameter	76.00	Mm
5	Burden	1.70	M
6	Spacing	1.70	M
7	Stemming height	1.60	M
8	Charged column	4.70	M
9	Number of holes	64.00	—
10	High explosive	1.56	Kg/Cartridge
11	Number of explosives used per hole	4.00	—
12	High explosive	400.00	Kg/Hole
13	ANFO	3.72	Kg/m
14	ANFO	719.00	Kg/Hole
15	Total weight of explosive	1119.00	Kg
16	Volume of rock	1165.25	m^3
17	**Powder factor**	**0.96**	**Kg/m^3**
18	Tonnage factor	2.70	T/m^3

Table 1.
Blast number one.

S/N	Parameters	Numeric value	
1	Explosive used	High Explosive + ANFO	
2	Drilling pattern	Staggered	
3	Hole depth	7.90	M
4	Diameter	76.00	Mm
5	Burden	1.80	M
6	Spacing	1.80	M
7	Stemming height	2.00	M
8	Charged column	5.90	M
9	Number of holes	88.00	—
10	High Explosive	1.56	Kg/Cartridge
11	Number of explosives used per hole	6.00	—
12	High explosive	825.00	Kg/Hole
13	ANFO	3.72	Kg/m
14	ANFO	1106.48	Kg/Hole
15	Total weight of explosive	1931.48	Kg
16	Volume of rock	2252.45	m^3
17	**Powder factor**	**0.86**	**Kg/m^3**
18	Tonnage factor	2.70	T/m^3

Table 2.
Blast number two.

S/N	Parameters	Numeric Value	
1	Explosive used	High Explosive + ANFO	
2	Drilling pattern	Staggered	
3	Hole depth	6.80	M
4	Diameter	76.00	Mm
5	Burden	1.90	M
6	Spacing	1.90	M
7	Stemming height	1.50	M
8	Charged column	5.30	M
9	Number of holes	74.00	—
10	High explosive	1.56	Kg/Cartridge
11	Number of explosives used per hole	4.00	—
12	High explosive	462.50	Kg/Hole
13	ANFO	3.72	Kg/m
14	ANFO	996.51	Kg/Hole

S/N	Parameters	Numeric Value	
15	Total weight of explosive	1459.01	Kg
16	Volume of rock	1816.55	m^3
17	**Powder factor**	**0.80**	**Kg/m^3**
18	Tonnage factor	2.70	T/m^3

Table 3.
Blast number three.

S/N	Parameters	Numeric value	
1	Explosive used	High Explosive + ANFO	
2	Drilling pattern	Staggered	
3	Hole depth	7.76	M
4	Diameter	165	Mm
5	Burden	4	M
6	Spacing	4	M
7	Stemming height	1.8	M
8	Charged column	5.5	M
9	Number of holes	92	—
10	High explosive	1.56	Kg/Cartridge
11	Number of explosives used per hole	4	—
12	High explosive	650	Kg/Hole
13	ANFO	18.73	Kg/m
14	ANFO	102.99	Kg/Hole
15	Total weight of explosive	10,137.00	Kg
16	Volume of rock	11,425.60	m^3
17	**Powder factor**	**0.89**	**Kg/m^3**
18	Tonnage factor	2.7	T/m^3

Table 4.
Blast number four.

S/N	Parameters	Numeric value	
1	Explosive used	High Explosive + ANFO	
2	Drilling pattern	Staggered	
3	Hole depth	13.28	M
4	Diameter	165	Mm
5	Burden	3	M
6	Spacing	4	M

S/N	Parameters	Numeric value	
7	Stemming height	1.3	M
8	Charged column	9.82	M
9	Number of holes	52	—
10	High explosive	1.56	Kg/Cartridge
11	Number of explosives used per hole	4	—
12	High explosive	408	Kg/Hole
13	ANFO	10.93	Kg/m
14	ANFO	107.34	Kg/Hole
15	Total weight of explosive	10,295.00	Kg
16	Volume of rock	8286.00	m^3
17	**Powder factor**	1.24	**Kg/m^3**
18	Tonnage factor	2.7	T/m^3

Table 5.
Blast number five.

S/N	Parameters	Numeric value	
1	Explosive used	High Explosive + ANFO	
2	Drilling pattern	Staggered	
3	Hole depth	14.41	M
4	Diameter	165	Mm
5	Burden	3	M
6	Spacing	4	M
7	Stemming height	1.8	M
8	Charged column	12.15	M
9	Number of holes	52	—
10	High explosive	1.56	Kg/Cartridge
11	Number of explosives used per hole	4	—
12	High explosive	760	Kg/Hole
13	ANFO	19.52	Kg/m
14	ANFO	107.34	Kg/Hole
15	Total weight of explosive	11,094.60	Kg
16	Volume of rock	8994.48	m^3
17	**Powder factor**	**1.23**	**Kg/m^3**
18	Tonnage factor	2.7	T/m^3

Table 6.
Blast number six.

S/N	Parameters	Numeric value	
1	Explosive used	High Explosive + ANFO	
2	Drilling pattern	Square	
3	Hole depth	16.44	M
4	Diameter	102	Mm
5	Burden	2.5	M
6	Spacing	2.5	M
7	Stemming height	1.52	M
8	Charged column	14.44	M
9	Number of holes	87	—
10	High explosive	1.56	Kg/Cartridge
11	Number of explosives used per hole	3	—
12	High explosive	4.68	Kg/Hole
13	ANFO	6.70	Kg/m
14	ANFO	91.53	Kg/Hole
15	Total weight of explosive	8237.95	Kg
16	Volume of rock	8938.06	m^3
17	**Powder factor**	**0.89**	**Kg/m^3**
18	Tonnage factor	2.7	T/m^3

Table 7.
Blast number seven.

S/N	Parameters	Numeric value	
1	Explosive used	High Explosive + ANFO	
2	Drilling pattern	Square	
3	Hole depth	12.07	M
4	Diameter	102	Mm
5	Burden	2.8	M
6	Spacing	2.8	M
7	Stemming height	1.8	M
8	Charged column	10.27	M
9	Number of holes	92	—
10	High explosive	1.56	Kg/Cartridge
11	Number of explosives used per hole	3	—
12	High explosive	4.68	Kg/Hole
13	ANFO	6.70	Kg/m
14	ANFO	60.30	Kg/Hole
15	Total weight of explosive	6046.37	Kg

S/N	Parameters	Numeric value	
16	Volume of rock	8707.57	m^3
17	**Powder factor**	**0.69**	**Kg/m^3**
18	Tonnage factor	2.7	T/m^3

Table 8.
Blast number eight.

S/N	Parameters	Numeric value	
1	Explosive used	High Explosive + ANFO	
2	Drilling pattern	Square	
3	Hole depth	15.59	M
4	Diameter	102	Mm
5	Burden	2.5	M
6	Spacing	2.5	M
7	Stemming height	1.5	M
8	Charged column	12.83	M
9	Number of holes	105	—
10	High explosive	1.56	Kg/Cartridge
11	Number of explosives used per hole	3	—
12	High explosive	4.68	Kg/Hole
13	ANFO	6.70	Kg/m
14	ANFO	60.30	Kg/Hole
15	Total weight of explosive	9176.62	Kg
16	Volume of rock	10,033.30	m^3
17	**Powder factor**	**0.91**	**Kg/m^3**
18	Tonnage factor	2.7	T/m^3

Table 9.
Blast number nine.

Sample	Maximum load (KN)	Specimen length (m)	Specimen width (m)	Cross sectional area (m^2)	Uniaxial compressive strength (MN/m^2)
Tutu Quarry	133.2	0.06	0.03	0.0018	74
F&P Quarry	145.8	0.06	0.03	0.0018	81
AAY Quarry	157.5	0.06	0.03	0.0018	87.5

Table 10.
Uniaxial compressive strength (UCS) MN/m^3.

charge, base charge, tonnage factor, quantity of material blasted and their corresponding powder factors adopted for rock fragmentation are presented for ten (10) blasting operations.

Details of the variability results are follows:

Drilling pattern adopted
Staggered.
Explosive type used
High explosive (Gelatin) + ANFO.
Hole depth
6.30 m: 16.44 m.
Hole diameter
76 mm, 102 and 165 mm.
Burden-spacing pattern
1.7 × 1.7: 2.8 × 2.8.
Stemming height
1.2 M: 2.7 m.
Charged column
4.0 M: 14.44 m.
Number of holes
58: 105.
Total weight of explosives
1119.00 kg – 11,094.60 kg.
Volume of rock blasted
1165.25 m^3–11,425.60 m^3.
Varied PF
0.77–1.24.
Constant TF
2.7.

2.5 Sample preparation and laboratory analysis

A circular saw with a diamond blade was used to cut the specimens to their final lengths. The surfaces were then ground after cutting in a grinding machine in order to achieve a high-quality surface for the axial loading. The measurement of the specimen dimensions was made with a sliding caliper and metre rule. Furthermore, the tolerances were checked by means of a dial indicator and a stone face plate. The specimen preparation was carried out in accordance with ASTM test procedure (ASTM, 39-71) and as adopted by Vandergrift and Schindler [23] in their experiment. The sample was cut using cutting machine to a dimension suitable for uniaxial compressive stress (UCS) test. The specimen was placed in horizontal direction but perpendicular to the direction of cutting edge of the blade. Then the vice was used to hold the specimen firmly to obtain a smooth surface as accurately as possible. The machine was switched on and the necessary shield applied. Water was allowed to lubricate the blade during the cutting process.

2.5.1 Procedure for uniaxial compressive strength test

The ASTM test procedure (39-71) was adopted. The specimen was placed in the ELE ADR 2000 compression machine. The load is continuously applied on the specimen until it failed. The failure mode was noted as well as the pressure or load at failure. The type of failure and the maximum load carried by the specimen were recorded. The unconfined UCS of the rock sample was obtained by dividing the maximum load carried by the cross-sectional area. Testing machine of standard recommended ASTM C 39-71 was used to load the squared sample until it failed.

2.5.2 Test specimens

Squared samples were used for this test. The four sides of each sample were ground flat, smooth and perpendicular to axis, that is they were parallel to each other 4 cm × 4 cm cube specimen were cut from block samples supplied (in the absence of core which are commonly used). The platens on the compression machine were altered to suit this configuration. The edges were cut to shape and smoothened by polishing them with carborundum powder.

3. Discussion of results

Findings from field studies and observations as shown in **Table 11** have revealed that the trend in the spread of GPS coordinates of granite samples collected indicates that the samples were taken at relatively close intervals. Results shown in **Tables 1–10** also indicate the various specifications of drilling and blasting parameters as used in the study site. From the tables it can be deduced that the stemming heights, burden spacing, column of charged holes and bulk density were varied as the depth of blast holes increases. It is also shown that the number of explosives charged per hole indicates an increase in the depth of hole drilled, thereby increasing more explosives consumption and equally implying that more volume of granite has been fragmented. **Table 10** shows uniaxial compressive strength (UCS) MN/m^3 of granite from the studied quarries, **Table 12** on the other hand shows the bulk densities of granite from the selected quarries. Also as apparently shown in **Table 13**, the mineralogy of the granitic constituents of granite is partly influenced by either a corresponding increase of decrease in PF. Hence, some sort of correlation seems to be found between granite mineralogy and PF. In addition, the PF for the nine (9) blasts was not the same, which means that it was varied at different times of the blast as shown in **Table 14**. For all the selected quarries, bigger blasthole diameter also varies with increased PF.

Results from this study have shown the level of relationship between the strength of rock and PF. For instance, it can be deduced from **Figures 2–4** that the Uniaxial Compressive Strength (UCS) of rock increases as the PF increases. This is also true in the reverse case as PF reduces. In the same vein, keeping UCS as a constant, PF increases as volume of blasted materials increases which is also true for the reverse case.

Sample	Latitude	Longitude
1	N26^0 51′7.2”	E007^0 59′ 71.4”
2	N26^{0}51’8.1”	E007^{0}59’75.5”
3	N26^0 51′13.3”	E007^{0}59’64.5”

Table 11.
Coordinates of samples collected.

Sample ID	Mass of specimen (g)	Height of specimen (m)	Width of specimen (m)	length of specimen (m)	volume of specimen (m^3)	Bulk density (Kg/m^3)
Tutu Quarry	2742.6	0.1	0.1	0.1	0.001	2743
F&P Quarry	2742.4	0.1	0.1	0.1	0.001	2742
AAY Quarry	2183.3	0.1	0.1	0.1	0.001	2183

Table 12.
Bulk density of selected quarries' granite samples.

Mineral	Sample 1 (%)	Sample 2 (%)	Sample 3 (%)	Average (%)
Quartz Syn	39	40	30	36.33
Microcline	0	0	28	9.33
Albeit	34.6	31	10	25.20
Orthoclase	25	17	16	19.33
Anorthite	0.3	10	6	5.43
Muscovite	0.1	0.8	2	0.97
Osumilite	0.9	0.7	8	3.20
TOTAL	99.9	99.5	100	99.80

Table 13.
Result of mineralogical analysis of samples.

S/N	Powder factor (Kg/m^3)	Volume of rock produced (m^3)	Location
1	0.96	1165.25	Tutu quarry, Kaduna
2	0.86	2252.45	Tutu quarry, Kaduna
3	0.80	1816.55	Tutu quarry, Kaduna
4	0.89	11,425.60	F&P Quarry, Abuja
5	1.24	8286.00	F&P Quarry, Abuja
6	1.23	11,094.60	F&P Quarry, Abuja
7	0.89	8938.06	AAY Quarry, Kano
8	0.69	10,033.30	AAY Quarry, Kano
9	0.91	8707.57	AAY Quarry, Kano

Table 14.
Powder factors and volume of rock produced.

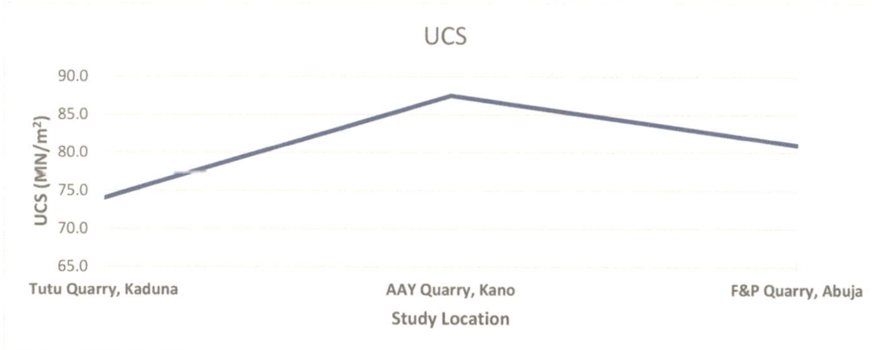

Figure 2.
UCS across the study locations.

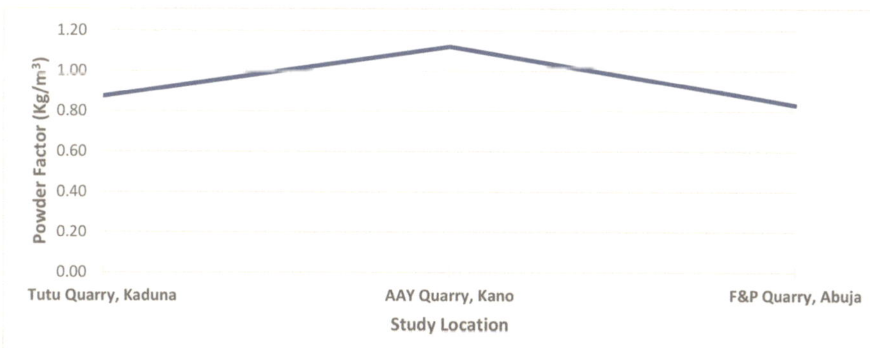

Figure 3.
Average powder factor (pf) across the study locations.

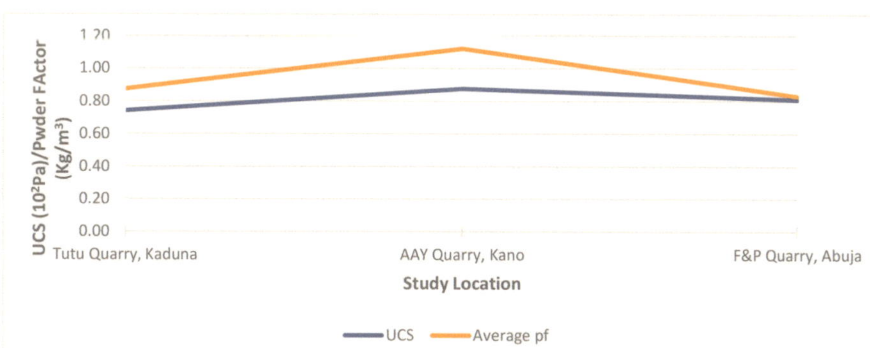

Figure 4.
Variability of the UCS and average powder factor (pf) across the study locations.

From the figures, it is evident that PF increases as the volume of blasted rock increases. Thus, more volume of rocks means higher PF to fragment the rock. Therefore, more explosives are charged in the holes to get the required results as the volume of rock increases. It can be deduced that to reasonably vary the PF of rock, the ratio of

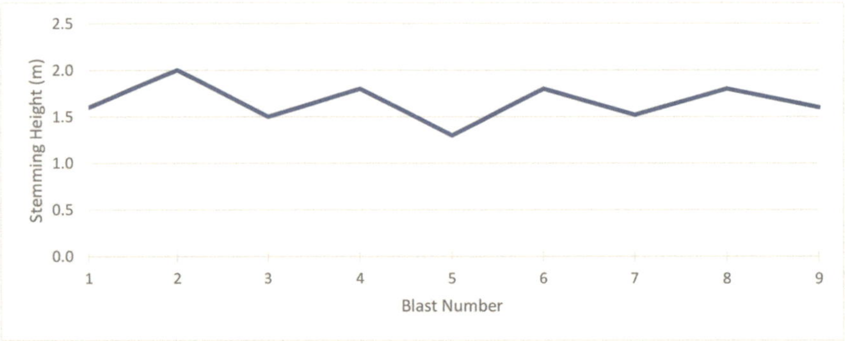

Figure 5.
Variability of the stemming height across the blasts.

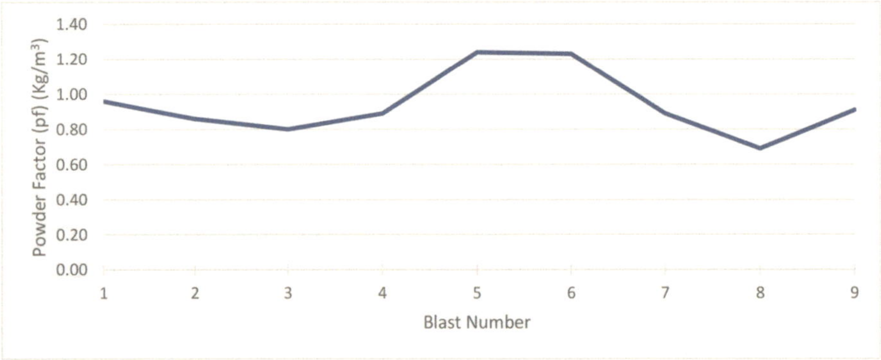

Figure 6.
Powder factor (PF) across the blasts.

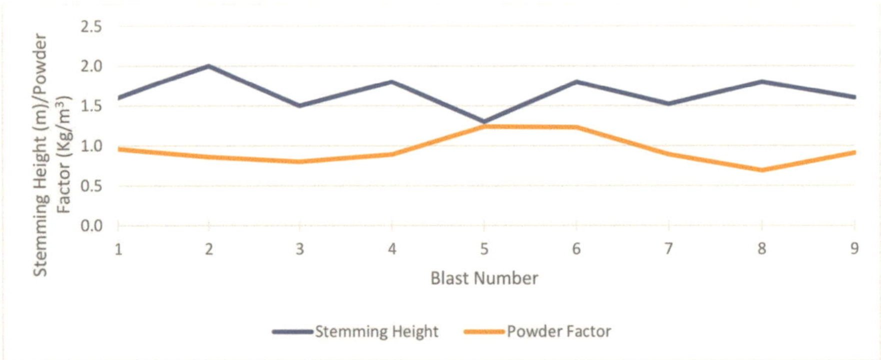

Figure 7.
Variability of the stemming height and powder factor (PF) across the blasts.

burden to spacing must be carefully selected with a view to increasing its productivity. Hence, PF variability becomes more effective with careful and staggered increment of burden and spacing.

As evident from **Figures 5–7**, more explosives are consumed when the volume of fragmented rock increases. Therefore, varied PF with carefully varied and selected drilling and blasting parameters are required for optimum blast and higher productivity.

4. Conclusions and recommendations

The PF used at the selected quarries has been examined for nine (9) blasts to ascertain its effects on the level of granite productivity in the three selected quarries. It can therefore be concluded that:

i. PF variability can lead to increase or decrease in the level of productivity of granite.

ii. Blasthole diameter varies with increase in PF of granite.

iii. PF selection and variability can lead to higher cost of productivity.

iv. The apparent correlation between PF and granite mineralogy can be a factor for consideration in blast design and the desired productivity.

v. The strength of granite influenced the PF selected at the selected quarries.

Also, the quantity of explosives for some of the sizes of the fragmented rock at the quarries was found to be moderate, while for some, it was optimal for the blast. Hence, it can be concluded from the observed blasts that the productivity of granite at a quarry can be improved for optimum economic benefit, if the PF is varied as appropriate for any blast design and the quality and properties of the explosives selected are adequate for the strength of the rock to be blasted.

Having observed and examined the effects of PF variability on the level of granite productivity, the following measures are hereby recommended for optimum rock fragmentation:

1. The strength of granite should be examined before selecting a suitable PF.

2. Since the PF variability has great impact on the productivity of granite, its selection must be guided by expertise and experience.

3. Economy and productivity of granite quarries have potentials for improvement when serious attention is given to the blast design and study of explosive characteristics and properties.

Author details

Luqman Kareem Salati[1]* and Moses Shola Adeyemo[2]

1 Kaduna Polytechnic, Kaduna, Nigeria

2 Vodem Technical and Engineering Consulting Ltd, Abuja, Nigeria

*Address all correspondence to: lksalati@gmail.com

References

[1] Akande JM, Lawal AI. Optimization of blasting parameters using regression models in Ratcon and NSCE granite quarries, Ibadan, Oyo state, Nigeria. Geomaterials. 2013;**3**:28-37

[2] Comakli R, Atici U, Dogangun E. Effect of Drilling and Blasting Performance on the Energy Consumption of a Jaw Crusher. Ankara, Turkey: 9th International Drilling and Blasting Symposium; 2017. pp. 187-191

[3] Eshun PA, Afum BO, Boakye A. Drill and blast performance evaluation at the Obra pit of Chirano gold mines ltd, Ghana. Ghana Mining Journal. 2016; **16**(2):28-35

[4] Salati LK, Mark GO. Optimum powder factor selection in blast holes at Dangote limestone quarry, Obajana, north-Central Nigeria. International Journal of Scientific and Engineering Research. 2020;**11**(1):1459-1476

[5] Mohamed F, Hafsaoui A, Talhi K, Menacer K. Study of the powder factor in surface bench blasting. Procedia Earth and Planetary Science. 2015;**15**:892-899

[6] Jethro MA, Shehu SA, Kayode TS. Effect of fragmentation on loading at Obajana cement company plc, Nigeria. International Journal of Scientific and Engineering Research. 2016;**7**(4): 608-620

[7] Liu M, Liu J, Zhen M, Zhao F, Xiao Z, Shan P, et al. A comprehensive evaluation method of bench blast performance in open-pit mine. Applied Sciences. 2020;**10**(5398):1-12

[8] Mkumbwa AG. Analysis of Drilling and Blasting Parameters to Achieve Optimum Rock Fragmentation at Songwe II Limestone Quarry at MCC, Unpublished BSc Thesis. University of Dodoma; 2017. p. 44

[9] Afum BO, Temeng VA. Reducing drill and blast cost through blast optimisation – A case study. Ghana Mining Journal. 2015;**15**(2):50-57

[10] Agyei G, Owusu-Tweneboah M. A comparative analysis of rock fragmentation using blast prediction results. Ghana Mining Journal. 2019; **19**(1):49-58

[11] Bhatawdekar RM, Mohamad ET, Singh TN, Armaghani D, J. Drilling and blasting improvement in aggregate quarry at Thailand – A case study. Journal of Mines, Metals and Fuels. 2019:357-362

[12] Vladimir M, Žarko K, Miodrag C, Jovana C. Economic cost analysis of drilling and blasting depend of drilling and blasting parameters at quarry "Dobrnja" near Banja Luka. Archives for Technical Sciences. 2015;**13**(1):35-41

[13] Agyei G, Nkrumah MO. A review on the prediction and assessment of powder factor in blast fragmentation. Nigerian Journal of Technology. 2021;**40**(2):275-283

[14] Salmi EF, Sellers EJ. A review of the methods to incorporate the geological and geotechnical characteristics of rock masses in Blastability assessments for selective blast design. Engineering Geology. 2021;**281**:1-37

[15] Abaje IB, Oladipo EO. Recent changes in the temperature and rainfall conditions over Kaduna state, Nigeria. Ghana Journal of Geography. 2019;**11**(2): 127-157

[16] Ati OF, Stigter CJ, Iguisi EO, Afolayan JO. Profile of rainfall change and variability in the northern Nigeria,

1953-2002. Research Journal of Environmental and Earth Sciences. 2009;**1**(2):58-63

[17] Janecke BB. Vegetation structure and spatial heterogeneity in the granite supersite. Kruger National Park', Koedoe. 2020;**62**(2):a1591

[18] Ladan SI. Forests and Forest reserves as security threats in northern Nigeria. European Scientific Journal. 2014; **10**(35):120-142

[19] Odekunle MO, Okhimamhe AA, Sanusi YA, Ojoye S. Local climate zone classification of the cites of Kaduna and FCT in Nigeria. Journal of Research in Forestry, Wildlife and Environment. 2019;**11**(4):27-40

[20] Hassan H, Waru SM, Bukar GA, Abdullahi KM. Groundwater potentials estimation of a basement terrain using pumping test data for parts of sanga local government area, Kaduna State, Northwestern Nigeria. Open Journal of Modern Hydrology. 2016;**6**:222-229

[21] Abere DV, Oyatogun GM, Ojo SA, Abubakar UBS, Otebe SI, Adejo OH, et al. Aggregate quarry at Gidan Tagwaye, Dutse local government area of Jigawa state. Aspects in Mining and Mineral Science. 2020;**5**(3):590-602

[22] Yusuf MF. Geology of Mahanga and its Environs Part of Sheet 103, Ikara SW, Ikara Local Government Area, Kaduna State, Unpublished BSc Thesis. Zaria: Ahmadu Bello University; 2019. p. 61

[23] Vandegrift D Jr, Schindler AK. The effect of test cylinder size on the compressive strength of Sulfur capped concrete specimens. In: Final Report on Highway Research Center. Alabama, PRO 2 - 13399: Auburn University; 2005. p. 84

Chapter 6

Physical and Mechanical Properties of Herrnholz Granite: An Ideal Experimental Material

Ying Li and Rui Wu

Abstract

Granite, as the most common plutonic rock of the Earth's crust and the most widely used paving block and building stone in industrial activities, has been widely employed in experimental investigations on its chemical composition, physical properties, and mechanical responses. This chapter focuses on the physical and mechanical properties of Herrnholz granite while emphasizing that it is an ideal experimental material for its homogeneity and fine-grained nature. Among the properties discussed here are density, porosity, pore size distribution, ultrasonic wave velocities, strength, fracture toughness, and hydroscopic/hygroscopic properties. Preliminary laboratory data sets to reveal relationships between the hygroscopic properties and mesoporous character of the Herrnholz granite as a result of water adsorption on internal fabric elements, such as pores, and microcracks.

Keywords: Herrnholz granite, fine-grained nature, homogeneity, mesoporous media, mechanical response, hygroscopic expansion

1. Introduction

Granite is the most common plutonic rock of the Earth's crust and has been used for a variety of rock engineering and geomechanical purposes, including underground disposal of radioactive waste (e.g., see [1–3]), cavern construction for liquid natural gas or liquid petroleum gas storage (e.g., see [4, 5]), and geothermal energy extraction from hot dry rock (e.g., see [6, 7]). To select a specific site for those geotechnical applications, geological (e.g., geometry, hydrology, and geochemistry), engineering (e.g., stress state, physical, and mechanical properties), and socioeconomic (e.g., seismicity, water, and land resources) conditions are typically taken into account [8–10]. Physical and mechanical properties of the host rock are normally evaluated at a later stage of the site selection process when all other factors are favorable. Granite has also been used as a construction stone, such as building facades, walls, sockets, and sculptures (see [11–14]), due to its abundance, petrophysical properties, durability, and textural uniformity. A detailed examination of the physical and mechanical properties of such granite helps to evaluate its long-term behavior in various building situations and environmental conditions.

The physical and mechanical properties, such as the bulk density, porosity, and uniaxial compressive strength, of a variety of granite have been reported (e.g., [7, 14–21]). In some cases, the ultrasonic velocity test [14, 22–25], the Schmidt hammer test [14, 25, 26], or the absorption test [14, 27–29] were performed, providing insights into the relationships between measured properties, for example, the relationship between UCS and porosity, between UCS and Schmidt rebound hardness, and between ultrasonic velocity and porosity. Note that these studies typically do not include microstructure characterization, such as grain size, grain shape, and pore size distribution, which indeed have been shown to be related to the mechanical and physical response [16, 30–32].

This chapter aims to provide an overall picture of Herrnholz granite by expanding on its microstructural characteristics and physicomechanical properties. For initial microstructure characterization, a series of techniques including optical petrographic and fluorescence (i.e., filling microcracks with a fluorescent dye) microscopy, mercury injection porosimetry, and nitrogen adsorption analysis were employed. Subsequent mechanical properties investigation involved a series of uniaxial compressive strength (UCS) tests and single edge notch bending (SENB) tests. Deformation and elasticity variation of Herrnholz granite in response to a range of relative humidity, referred to as the hygroscopic properties, and to the progressive water imbibition, referred to as the hydroscopic properties, were evaluated under controlled climatic conditions using a unique combination of on-specimen strain, applied load and displacement, and digital image correlation (DIC).

2. Granite studied

Granite studied in this chapter was sourced from the Herrnholz quarry (HQ), east of the Hauzenberg pluton (HP), NNW of Munich, in Germany (**Figure 1a**). The Hauzenberg pluton ($\sim$ 100 km^2 in area), together with the Kristallgranite ($\sim$ 400 km^2 in area) and the Fuerstenstein ($\sim$ 100 km^2 in area) plutons, are commonly located between the Danubian fault and the Bavarian Lode shear zone (**Figure 1b**), and represent the westernmost part of the South Bohemian Massif. Granite in the Hauzenberg pluton was formed during the Variscan orogeny and intruded late-to post-kinematically at $\sim$ 320 Ma (e.g., [33, 34]) within the Moldanubian part of the southwestern Bohemian Massif. Rapid cooling (100°C/Ma $\sim$ for 2–3 Ma) and a single phase of exhumation under relatively consistent tectonic conditions have produced homogeneous granite with minimal to no ductile tectonic overprint, which makes it a popular building stone in the region.

Two hundred samples were cut from a single 0.6 m^3 block (**Figure 1d**) of Herrnholz granite for a range of geomechanical laboratory testing. The block showed little to no discoloration or staining, indicating its unweathered nature. All samples were oriented relative to the principle splitting directions, the 'rift', 'grain', and 'hardway'. The rift is the plane along which the granite cleaves with the greatest ease, followed by the grain, and lastly by the hardway. The easiest splitting direction is most likely due to parallel micro-cracks in the homogeneous fine-grained granite, the grain may be determined by inherited mineralogical characteristics (e.g., cleavage of feldspar and mica), or tectonic fractures, and the hardway may be a direction at right angles to the other two.

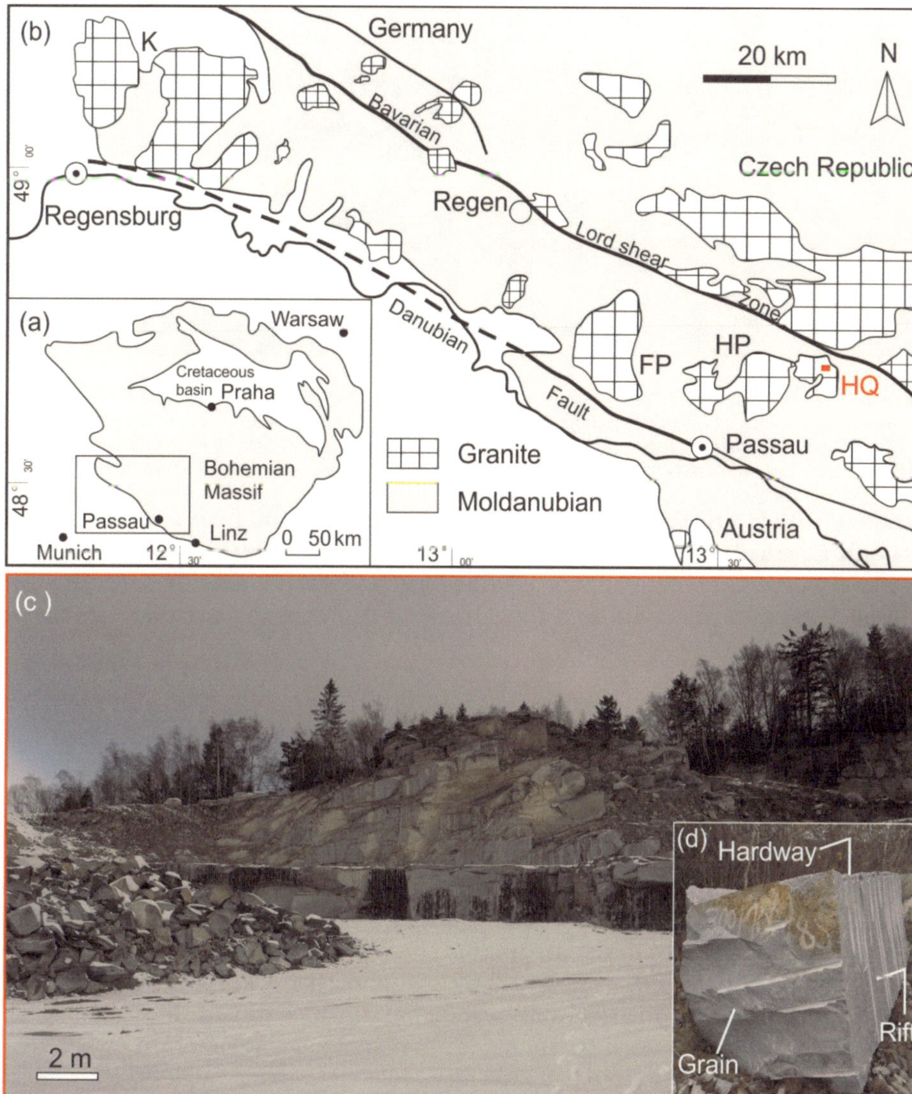

Figure 1.
(a) Map showing the location of the Bohemian Massif; (b) Simplified geological map (modified after Refs. [33, 34]) showing the location of Herrnholz quarry (HQ) with respect to the distribution of three largest granite intrusive plutons in the westernmost part of the South Bohemian Massif: FP, Fuerstenstein Pluton; HP, Hauzenberg Pluton; K, Kristallgranite. (c) Photo of the Herrnholz quarry front; and (d) photo of the selected block for samples preparation, showing rift, grain, and hardway planes.

A suit of 35 mm × 22 mm × 15 mm Herrnholz granite prisms was prepared for thin sections (35 mm × 22 mm, ~ 30 ± 5 μm thick) observation. Biotite grains can be distinguished from others under plane-polarized light by their brownish appearance (**Figure 2a**), while muscovite crystals are distinctive in crossed-polarized light by their mottled appearance with rainbow pattern (**Figure 2b**). Quartz can be distinguished from feldspar because it is generally clear and lacks visible twinnage or cleavage, despite all being shades of dark gray through to white in crossed-polarized light. These observations indicate granitic mineralogical assemblage of our selected

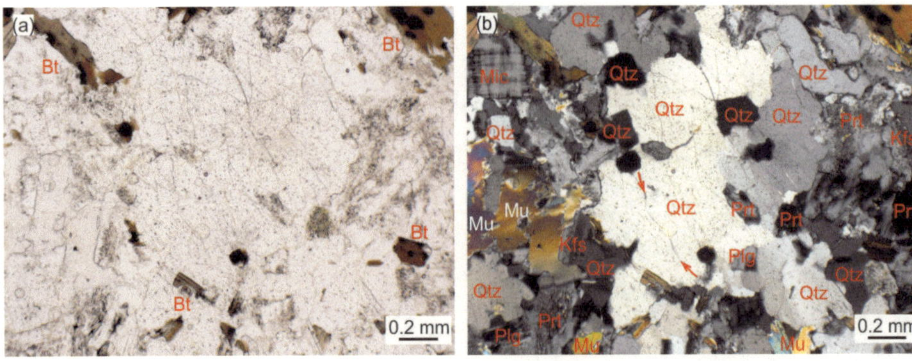

Figure 2.
Petrographic thin section of Herrnholz granite (a) in plane-polarized light, showing biotite (Bt) in brown; (b) in crossed-polarized light, showing interlocking quartz (Qtz), K-feldspars (Kfl) in form of microcline, perthite (Prt), and plagioclase (Plg), all in shades of dark gray through to white, and muscovite (Mu) with rainbow patterns.

specimen, with interlocking quartz (50% in area fraction) and feldspars (38% in area fraction), together with a small portion of mica (11% in area fraction). The grain size, calculated as average length of each grain's longer and shorter axes, is generally between 0.03 and 1 mm, with a mean diameter of 0.23 mm and a standard deviation of 0.13 mm.

3. Methods

3.1 Microstructure observation

Pore spaces (e.g., pores and cracks) within a thin section are hardly detected by microscope examination when not highlighted by iron or other filling minerals (e.g., [15]). A combination of optical petrographic and fluorescence microscopy was therefore employed to characterize the distribution of pore spaces in Herrnholz granite. Thin sections were dyed with a fluorescent pigment, which caused throughgoing pores and cracks to glow neon-green under ultraviolet light. We merged images obtained with crossed-polarized light and ultraviolet light into a single 3.2 Mpx mosaic (**Figure 3**), over which various cracks, for example, grain boundary cracks, intragranular cracks, and intergranular cracks connecting grain boundaries to the inside of grain, were able to be detected. In general, quartz crystals most commonly contain single intragranular or intergranular cracks, while feldspar crystals commonly contain a population of cleavage-parallel intergranular cracks.

Total porosity (ϕ_t), defined as the fraction of bulk volume occupied by the total pore space, can be determined from the difference between the bulk and grain density. Bulk densities (ρ_b) of two oven-dried Herrnholz granite prisms ($25 \text{ mm} \times 25 \text{ mm} \times 40 \text{ mm}$) at a temperature of 80°C for three days ($< 0.1\%$ mass change in 24 h) were calculated as the ratio of oven-dry mass to bulk volume. The masses of these samples were directly measured using an analytical balance (accuracy of 0.001 g), and the volume was determined from vernier caliper (accuracy of 0.01 mm) measured dimensions. Grain densities $\left(\rho_g\right)$ of the same samples were measured using a helium pycnometer (model: AccuPyc II196, the accuracy of 0.02 m^3 in grain volume). After placing the oven-dried sample in a sample cell of known

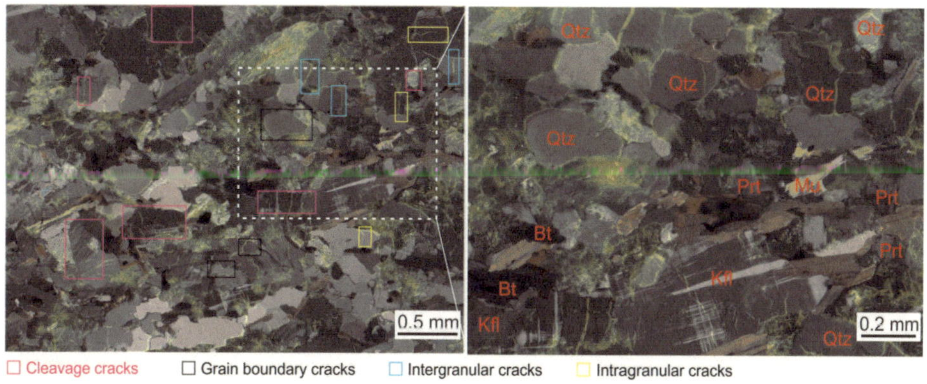

Figure 3.
Superimposed micromosaic obtained with crossed-polarized light and ultraviolet light, indicating regions of (i) cleavage cracks: in straight and parallel patterns within a grain; (ii) grain boundary cracks: along grain boundaries; (iii) intergranular cracks: connecting grain boundaries to the inside of a grain; and (iv) intragranular cracks: within a grain in random or parallel distributed sets (Qtz: quartz; Kfl: K-feldspar; Prt: perthite; Bt: biotite; Mu:muscovite).

volume at an initial pressure, helium gas was admitted to fill the sample cell and the resulting pressure was measured. The grain volume can be calculated from the two measured pressures, the known volume of the sample cell, and the added helium gas using the ideal gas law.

Other pore space properties, including accessible porosity, pore size distribution, and specific surface area of the Herrnholz granite were quantified using a combination of mercury injection porosimetry and nitrogen adsorption analysis. A suite of seven 20 mm × 6.5 mm × 6.5 mm Herrnholz granite prisms was prepared for mercury porosimetry. From measured intrusive volume of mercury under controlled pressures, mercury injection porosimetry can provide information on pore volume or porosity, as well as a wide range of pore throat size (typically from 3 to 10 nm up to microns). The nitrogen adsorption method can evaluate the size of mercury-inaccessible small pores by measuring the amount of adsorbate at a series of relative pressures. To this end, two intact 40 mm × 10.5 mm × 10.5 mm samples were prepared for nitrogen adsorption analysis using an automated gas sorption analyzer, Autosorb iQ^{TM}. The existing instrument software AsiQwin™ computed connected surface area and pore size distribution based on the Brunauer-Emmett-Teller (BET) [35] and BarretJoyner-Hallenda [36] model, respectively. The details of these models have been discussed in the mentioned papers; hence, they were not explained here.

3.2 3D ultrasonic tomography

3D ultrasonic tomography on three cuboidal (160 mm × 160 mm × 160 mm) specimens of Herrnholz granite was performed under ambient conditions. Three dimensions of the tested specimens relative to the quarry fabric, that is, rift, grain, and hardway (see **Figure 1d**) were denoted as *G1*, *G2*, and *G3*, respectively. **Figure 4** presents a schematic representation of the ultrasonic tomography setup along the *G1* direction (as an example), in which an array of nine in-house piezoelectric (PZT) transmitter (model: PCT-MCX) [37] and nine passive PZT receivers (model: KRNBB-PC) [38] were mounted on the top and bottom surfaces of the granite cube,

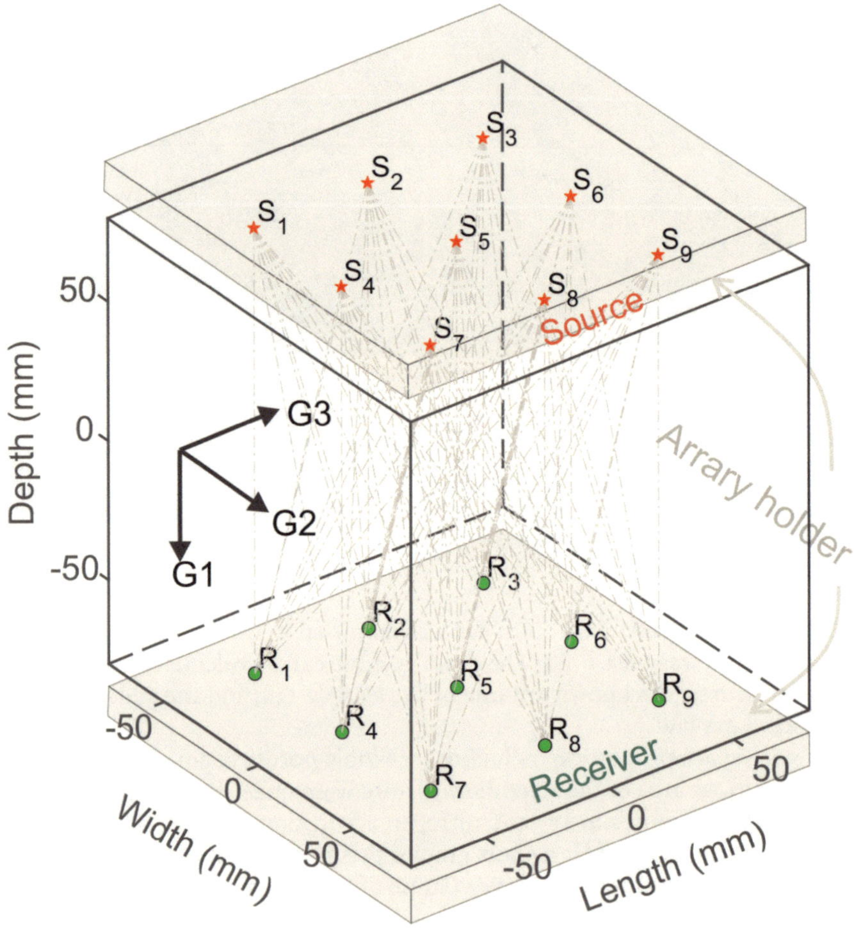

Figure 4.
Schematic representation of ultrasonic tomography setup. G1, G2, and G3 denote the three dimensions relative to the three principle splitting directions, and S and R indicate the Source and Receiver, respectively.

respectively, through two aluminum array holders (gray blocks, see **Figure 4**). A pulsing unit was used to apply a 300 V impulse source, with a duration of 1 μs, to the PZT transmitters. Each transmitter emitted an impulse source in a sequential manner, and the waveforms on nine receivers were recorded by the data acquisition system at a sampling rate of 20 MHz and 16-bit resolution.

Each specimen was modeled as 6 × 6 × 6 cubic elements, each having dimensions of 26.7 mm × 26.7 mm × 26.7 mm, for tomography. As indicated by dashed lines in **Figure 4**, there were 81 straight ray paths (assuming wave propagation by straight rays) that sample most of the elements between the transmitter and receiver planes (i.e., *G*1 planes). The time intervals for the first P-wave arrivals from the transmitters to the receivers and the distance of each transmitter–receiver pair were stored in two 9 × 9 arrays to derive the P-wave velocity array. P-wave velocity structure was derived at the center of each cubic elements using the Moore-Penrose pseudoinverse. More detail and the mathematical description of this inversion problem have been described in Ref. [39]; hence, they were not explained here.

3.3 Rock mechanics laboratory testing

In order to determine elastic and brittle properties of Herrnholz granite, a range of uniaxial compressive strength and single-edge notch three-point bending (also known as three-point bending) tests were performed with a custom static loading frame (see details in Ref. [40]) at the Rock Physics and Mechanics Laboratory, ETH Zurich. With the loading frame, the applied load and piston displacement are directly measured with a load cell and displacement transducer, respectively. Surface strains of a tested sample can be tracked using a range of extensometry products connected to the test machine controller. These include a radial chain extensometer (Epsilon Model 3544) and a four-point averaging longitudinal extensometer (Epsilon Model 3442RA1), which are often employed together in UCS tests to measure radial and axial strains, respectively, of a cylindrical sample as it is compressed (see **Figure 5a**), and a crack mouth opening displacement gauge (Epsilon Model 3541) that can be used directly on a SENB specimen with the knife edges glued to the test specimen or, alternately, with bolt-on knife edges mounted on the test specimen. In order to avoid effects associated with creeping of the adhesives or localized damage during bolt-hole preparation, the crack mouth opening displacement gauge in this study is clamped by knife edges manufactured within an L-shape titanium holders, of which the 8 mm-deep leg is embedded in the notch (> 8 mm deep) of a SENB specimen (see **Figure 5b**).

3.3.1 Uniaxial compressive strength tests

UCS tests were performed on 10 cylinders (50 mm in diameter, 140 mm in height) under ambient conditions to determine elastic properties, for example, Young's modulus (E) and Poisson's ratio (v), and brittle properties that include the crack initiation

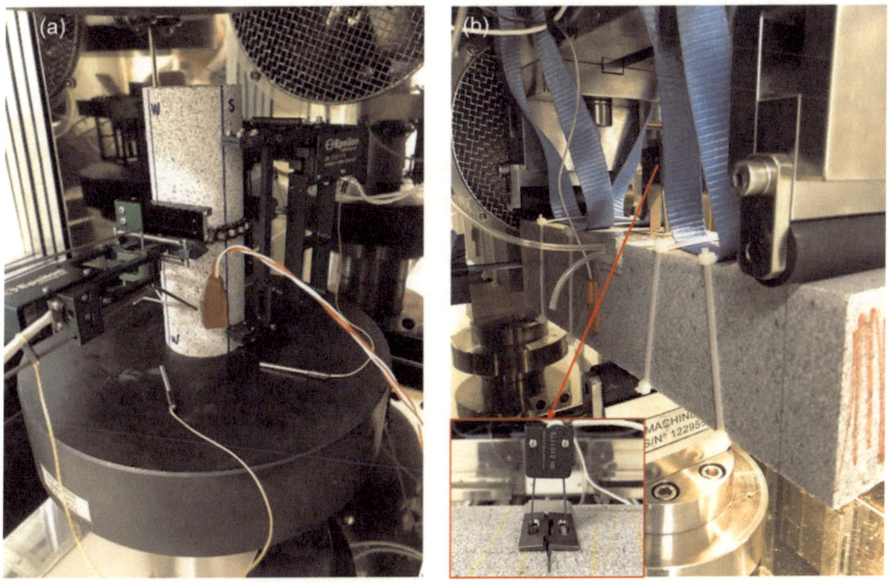

Figure 5.
Setup of uniaxial compressive strength (UCS) and single edge notch three-point bending (SENB) tests with a custom static loading frame at the Rock Physics and Mechanics Laboratory, ETH Zurich. (a) A UCS specimen was mechanically mounted with longitudinal and circumferential extensometers; (b) A SENB specimen was mounted with a knife edges-clamped crack mouth displacement gauge.

(CI) threshold, critical damage (CD) threshold, and UCS of Herrnholz granite. Every specimen was cored in the same direction and ground into two parallel end surfaces. The specimens were loaded at a constant displacement (piston) rate (4 mm/min) to failure, with load and displacement (incl. piston and sample) being logged over the failure process. To minimize the potential of damaging the displacement transducers, the failure of each specimen was arrested by a servo-controlled break detection mechanism that instantaneously unloaded the specimen when the load dropped more than 0.02% of the peak.

Stress–strain curves developed during the failure process were used to determine both the elastic and brittle properties. Young's modulus was determined as the slope of the linear portion of each axial stress–axial strain (σ_a vs. ϵ_a) curve, basically between 20% and 50% of the average UCS. Poisson's ratio was determined by linear regression of the radial strain–axial strain (ϵ_r vs. ϵ_a) curve over the same stress interval. The CI threshold represents the axial stress at which new cracks begin to form and can be determined using the crack volumetric strain reversal method [41, 42]. In this method, the crack volumetric strain (ϵ_{CV}) is calculated using the following formula:

$$\epsilon_{CV} = \epsilon_{vol} - \epsilon_{EV} \tag{1}$$

where ϵ_{vol} is the volumetric strain, assumed to be $\epsilon_a + 2\epsilon_r$,ϵ_{EV} is the elastic volumetric strain and can be calculated by:

$$\epsilon_{EV} = \frac{1-2v}{E}\sigma_a \tag{2}$$

The CD threshold corresponds to the start of unstable crack growth and can be determined by the reversal point in the volumetric strain-axial stress (ϵ_{vol} vs. σ_a) curve.

3.3.2 Single edge notch bending tests

SENB tests were performed on 4 Herrnholz granite beams under ambient conditions to determine their fracture toughness. Every beam is 400 mm long, with a span length between the bearings of 360 mm, a cross-section of 90 × 90 mm, and a 3 mm wide, 9 mm deep saw-cut notch in the outer bending radius. Two of the samples were loaded to failure under load-point displacement (piston) control at a rate of 1 μm/s. The other two samples were subjected to staged loading increases (ranging from 50 % to 98 % of the predetermined peak load) with load-point displacement maintained for up to 30 min between each load stage, aiming to gain insights into the time-dependent behavior of the Herrnholz granite. The fracture toughness of Herrnholz granite can be determined from the measured peak load using the following formula:

$$K_{IC} = \frac{F}{w\sqrt{h}}\left[\frac{3\frac{l}{h}\sqrt{\frac{a}{h}}}{2\left(1+2\frac{a}{h}\right)\left(1-\frac{a}{h}\right)^{\frac{3}{2}}}\left[1.99-\frac{a}{h}\left(1-\frac{a}{h}\right)\left[2.15-3.99\frac{a}{h}+2.7\left(\frac{a}{h}\right)^2\right]\right]\right] \tag{3}$$

where F is the measured peak load, l is the span length, w and h are the sample width and height, respectively, and a is the initial depth of the notch.

3.4 Water imbibition and adsorption tests

Increases in the water content of intact rock, either through exposure to high ambient humidity, or the addition of liquid water, have been shown to increase strains [27–29, 43] and change mechanical properties [44–49]. To differentiate the conditions that lead to these variations, we refer to them as hygroscopic (when related to humidity changes) and hydroscopic (when related to water immersion and/or imbibition) properties. The hydroscopic and hygroscopic properties of Herrnholz granite were determined through water imbibition experiments on two free-standing 115 × 65 × 35 mm granite prisms, and water vapor adsorption experiments on a single granite cylinder (50 mm in diameter and 140 mm in length) subjected to unconfined compression, respectively.

3.4.1 Liquid water imbibition tests

Water imbibition tests on two 115 × 65 × 35 mm granite prisms were performed with distilled water as the wetting fluid proceeded from the top to the bottom of each specimen as a combined result of gravity and capillary effect. The test arrangement is schematically presented in **Figure 6**, where water was introduced to the upper sample surface through two pieces of 65 mm wide filter paper folded with one end covering the sample surface and the other immersed in a distilled water reservoir ~ 15 mm above the specimen. Aluminum blocks were placed on the filter paper to ensure a positive contact surface between the saturated filter paper and the sample. In order to track the surface deformation of the specimens undergoing gradual wetting, time-lapse photographs were acquired using a combination of Sony Alpha A7RII digital camera and Sony FE 24–105 mm F4 G OSS zoom lens with the sensor plane parallel to the vertical plane (65 × 115 mm) of the tested sample. Digital Image Correlation (DIC) analysis of hundreds of photographs taken over 16–24 h was undertaken in Ncorr software [50]. The details of this software have been discussed in the mentioned papers; hence, they were not explained here.

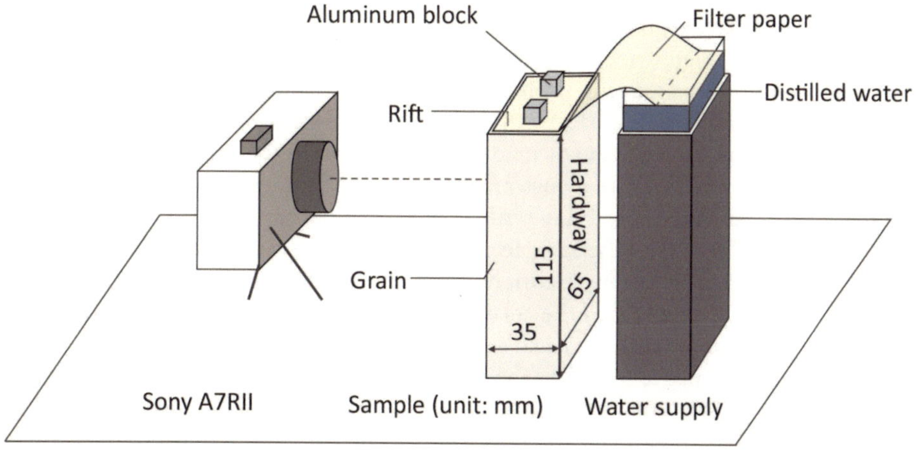

Figure 6.
Schematic diagram of the capillary imbibition experiment on a free-standing 115 × 65 × 35 mm granite prism.

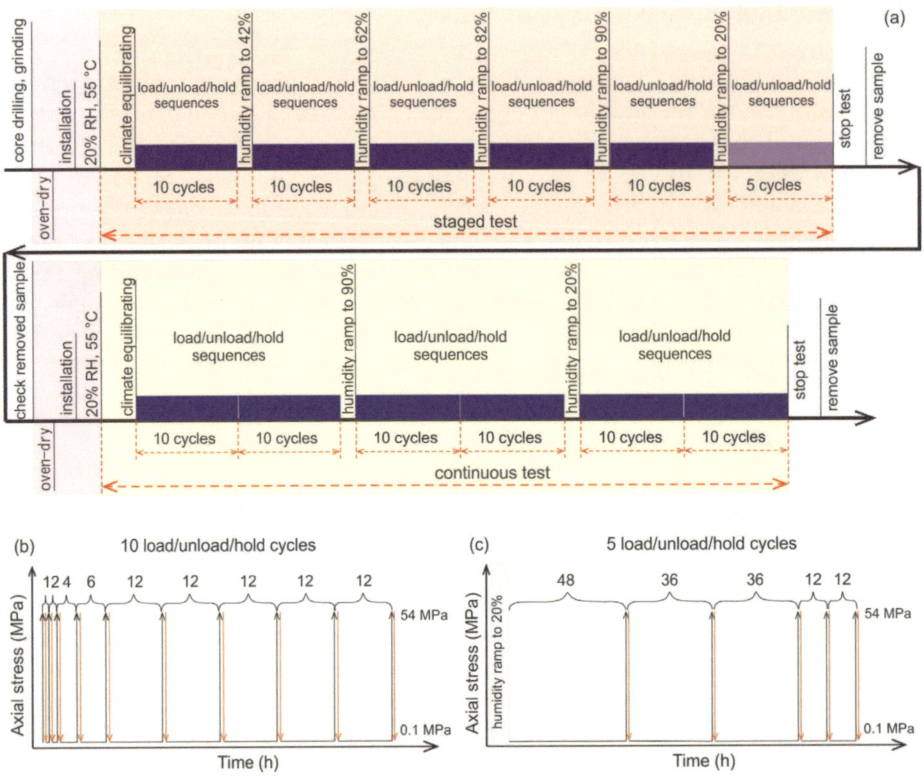

Figure 7.
Flowchart of the "staged" and "continuous" uniaxial compression tests (a), with panels (b and c) illustrating the 10 load/unload/hold cycles and 5 load/unload/hold cycles, respectively.

3.4.2 Water vapor adsorption tests

To derive hygroscopic properties of Herrnholz granite, two uniaxial compression tests on a single Herrnholz granite cylinder (50 mm in diameter and 140 mm in length) exposed to "staged" and "continuous" humidity variation were performed. The first "staged" test employed a humidity stepping protocol, before the sample was removed, checked, oven dried, and prepared for the second test, which employed a "continuous" humidity stepping protocol (see flowchart in **Figure 7a**). The sample was dried in an oven at 80°C for 3 days prior to each test to ensure identical initial moisture conditions and to prevent inadvertent damage to the microstructure. The sample was then transferred to a custom static loading frame with integrated climate chambers [40] and loaded for 1 day with a constant axial stress of 0.1 MPa, 20% humidity, and 55°C to allow the sample to equilibrate. During the staged test, relative humidity was ramped in a stepwise manner from 20 to 42, 62, 82, and 90% before returning to 20%. We increased the relative humidity from 20% to 90% in the continuous test before returning to 20%. In both tests, the axial stress increased at a rate of 1 MPa/s from 0.1 to 54 MPa (the predetermined CI threshold), then decreased at the same rate and remained constant at 0.1 MPa for 1 and 48 h (**Figure 7b** and **c**). Throughout the hold stages, cumulative strains were observed while axial stress was held constant at 0.1 MPa. During the loading stages, axial stress-strain relationships were used to derive elastic properties at each humidity level.

4. Results and discussion

4.1 Density, porosity, and pore size distribution

The mass and volume ratios of three tested specimens indicate an average bulk density of 2.61 g/cm^3, and grain density of 2.664 g/cm^3. Their uncertainties $\left(\frac{\delta\rho}{\rho}\right)$ resulted from the uncertainties in mass $\left(\frac{\delta m}{m}\right)$ and volume $\left(\frac{\delta V}{V}\right)$ measurements and can be evaluated by:

$$\frac{\delta\rho}{\rho} = \sqrt{\left(\frac{\delta V}{V}\right)^2 + \left(\frac{\delta m}{m}\right)^2}. \tag{4}$$

This provides an uncertainty of 0.08% or 0.002 g/cm^3 in the bulk density, and 0.10% or 0.003 g/cm^3 in the grain density.

Total porosity (ϕ_t) was derived from the difference between the grain and bulk density, that is, $1 - \rho_b/\rho_s$, as 1.9%. Its uncertainty $\left(\frac{\delta\phi_t}{\phi_t}\right)$ was given by

$$\frac{\delta\phi_t}{\phi_t} = \frac{1}{\rho_g - \rho_b}\sqrt{\left(\frac{\rho_b}{\rho_g}\delta\rho_s\right)^2 + (\delta\rho_b)^2} \tag{5}$$

and calculated as 0.2%.

The combination of mercury injection porosimetry and nitrogen adsorption analysis provides a wide range of pore size distributions over 3 nm to 100μm (**Figure 8**). The conversion of mercury intrusion pressure to the corresponding pore throat size (i.e., the Washburn equation, 1921) indicates pore throat sizes ranging from 10 nm to 100 μm (see **Figure 8a**). Nitrogen adsorption analysis shows that pores with a diameter smaller than 10 nm account for 80% of the specific surface area, which is averaged at 0.46 m^2/g (see **Figure 8b**) based on the BET model.

The intrusive volume of mercury ranged from 2.63 to 6.17 mm^3/g (mean at 4.22 mm^3/g) at pressures up to $\sim$ 400MPa (see **Figure 8a**), suggesting an average mercury-accessible porosity of 1.15%. The nitrogen adsorption analysis provides a pore volume of sim 0.5 mm^3/g (or a porosity of 0.14%) over a pore diameter range of 3–10 nm, which, when compared to the pore volume (4.22 mm^3/g) or porosity (1.15%) accessible by mercury, implies that only 1/12 of the small pores are accessible to nitrogen. The cumulative pore volume measured from mercury injection porosimetry ($\sim$ 1.2 mm^3/g) is slightly greater than that from the nitrogen adsorption analysis ($\sim$ 0.8 mm^3/g) over the overlapped diameter range of the two techniques, that is, 10 – 100 nm, indicating a greater differential of pore volume with respect to pore diameter. This slight inconsistency may be due to the nonuniform pore geometry (e.g., wedge-shaped pore) of the analyzed sample [51].

4.2 Ultrasonic wave velocities

3D ultrasonic tomography shows a nearly isotropic P-wave velocity along three directions and a uniform P-wave velocity structure in each plane. **Figure 9a** as an

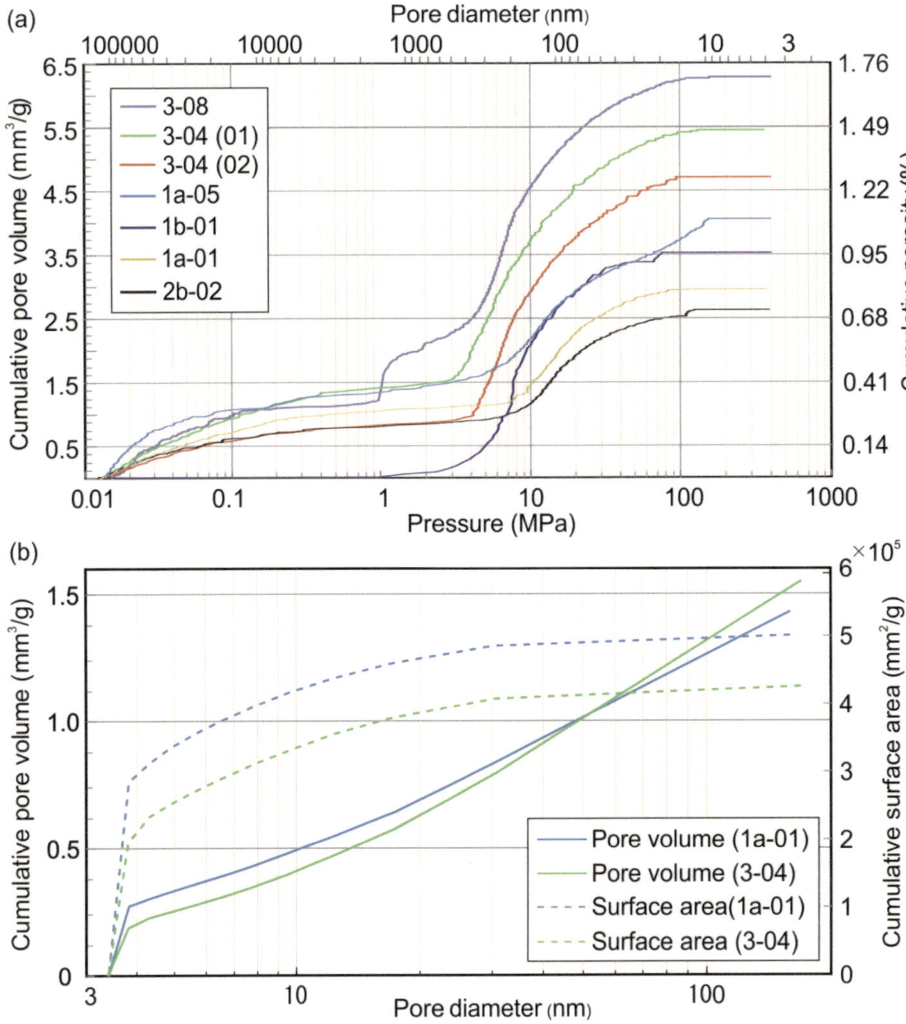

Figure 8.
Comparison of nitrogen adsorption analysis and mercury injection porosimetry results for pore size distribution. (a) Cumulative pore volume and mercury-accessible porosity as a function of pore throat diameter and applied pressure by mercury porosimetry. (b) Cumulative pore volume and surface area as a function of pore diameter during nitrogen adsorption.

example demonstrates the structure of P-wave velocity along the *G1* direction, giving a mean velocity of 3981 m/s, with standard deviation of 68.5 m/s among 64 elements. Note that P-wave velocities of the outermost elements on the *G2* and *G3* surfaces were omitted given the lack of effective coverage of straight rays. P-wave velocities in the *G2* and *G3* direction were similarly characterized, giving 3977 ± 60, and 3988 ± 64 m/s, respectively. Single-peaked normal (or Gaussian) distributions of P-wave velocities in three directions were shown in the probability density functions of **Figure 9b**, with very close peak values (cf. 3997 m/s, 3995 m/s, and 3995 m/s) in three directions. We therefore conclude isotropic wave propagation along three principle splitting directions, and a high degree of homogeneity in P-wave velocity over each plane of the Herrnholz granite. Repeated analysis on the other two specimens was carried on, and the overlapping probability histograms of P-wave velocity

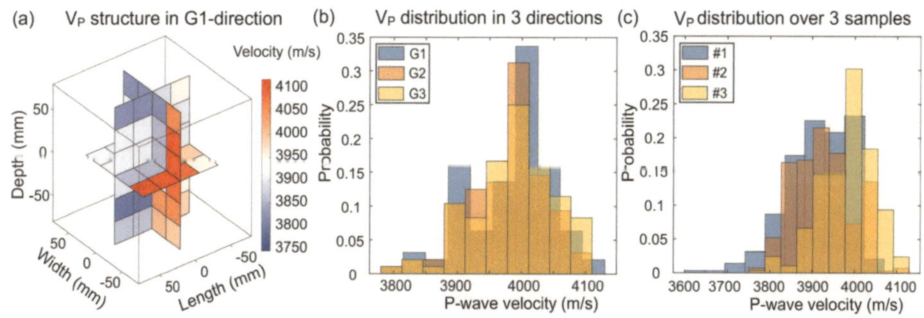

Figure 9.
P-wave velocity tomography of three cuboidal (160 mm × 160 mm × 160 mm) specimens of Herrnholz granite. (a) P-wave velocity structure in the G1 direction; (b) P-wave velocity histograms along the G1, G2, and G3 directions of one specimen. (c) P-wave velocity histograms for three tested specimens.

distribution were presented in **Figure 9c**, showing average P-wave velocities of $3914 \pm 74, 3925 \pm 71$, and 3982 ± 64 m/s, respectively, of three tested specimens.

4.3 Elastic properties, brittle properties, and progressive failure characteristics

Axial stress–strain curves (**Figure 10**) derived from ten UCS tests indicate a consistent brittle failure process among ten tested Herrnholz granite cylinders: (i) closure of preexisting cracks prior to a linear stress response; (ii) linear elastic behavior corresponding to a linear portion of the stress-strain curve; (iii) stable crack growth over the CI and CD interval; and (iv) unstable crack growth over the CD threshold, which leads to failure at the peak stress (i.e., the UCS). The measured UCS ranges between 127 and 158 MPa, with a mean value of 143 MPa, and a standard deviation of 12 MPa. Axial stress-axial strain curves (**Figure 10a**) present a consistent linear portion between 30 and 35% of the UCS. Young's modulus and Poisson's ratio were calculated by deriving the linear stress–strain relationship (see details in Section 3.3.1)

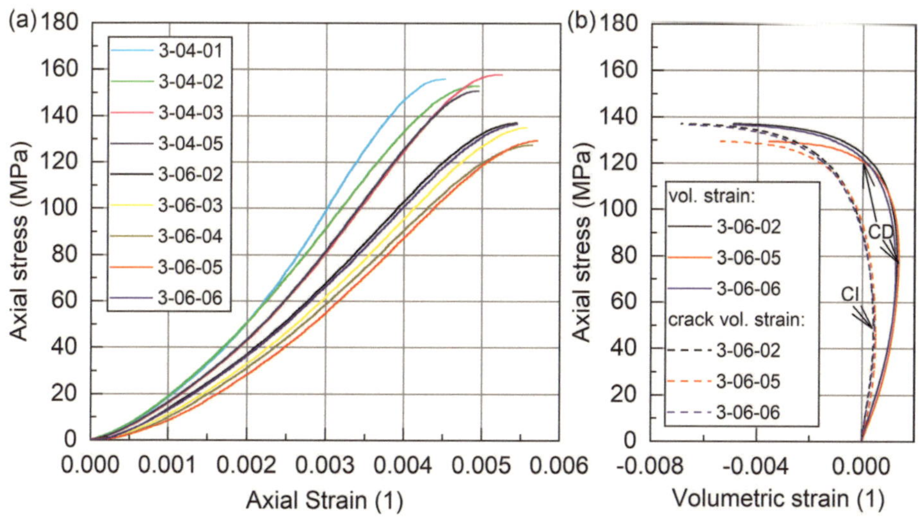

Figure 10.
Axial stress-strain curves derived from uniaxial compression tests on herrnholz granite cylinders in ambient conditions. (a) Axial stress-axial strain curves overloading; (b) Axial stress-volumetric strain curves overloading. CI: crack initiation threshold; CD: crack damage threshold.

over this interval, indicating a mean Young's modulus of 35 GPa, and Poisson's ratio of 0.28 under ambient environmental conditions.

Three of the tested samples were instrumented with a radial chain extensometer (see **Figure 5a**), which allowed for the determination of radial strain, and thus the volumetric strain and crack volumetric strain. The crack volumetric strain reversal method (see in Section 3.3.1) indicates an average CI threshold of 47 ± 1.2MPa (**Figure 10b**), $\sim$ 35% of the UCS, consistent with existing measurements typically between 30 and 35% of the UCS [42, 52–55]. The CD threshold is estimated to fall between 58% UCS and 92% UCS, corresponding to the volumetric strain reversal point [53, 56, 57] and the dilation point (transition of volumetric strain from positive to negative), respectively.

SENB tests demonstrate an exceptionally consistent failure load of $14.54\ \text{kN} \pm 0.18\text{kN}$, suggesting a (theoretical) average fracture toughness of $1.82 \pm 0.02\ \text{MPa} \cdot \text{m}^{1/2}$ following Eq. (3). Progressive failure characteristics during the load relaxation phase of one staged test (see test description in Section 3.3.2) were observed through a progressive increase in crack mouth opening displacement when the piston displacement was held at 98% of the predetermined peak load. This, however, requires to be supported by additional tests (e.g., [58]).

4.4 Hydroscopic expansion

Horizontal and vertical expansion in association with water imbibition were observed in two tested Herrnholz granite specimens. **Figure 11** demonstrates the evolution of vertical strains of one tested sample (as an example) during water imbibition. Adopting a compression negative convention in reporting strains, the vertical strains are extensional (**Figure 11**) upon wetting and progress to more than 30 mm below the top surface at the end of wetting (**Figure 11f**).

Heterogeneities observed in the vertical strain field (see **Figure 11**) indicate imbibition occurs along preferential capillary conduits that appear to be oriented parallel to the micro-crack fabric. Although laboratory studies have previously found little variability in the initial (24 h) capillarity of granite containing a population of preferentially oriented micro-cracks, longer-term (38 days) water uptake can be up to 20% greater in the grain-parallel orientation than the rift parallel [13].

Calculated mean linear strain assumed to be $(\epsilon_{\text{vertical}} + 2\epsilon_{\text{horizontal}})/3$, progressively increased with water imbibition in both samples, up to 4.7×10^4, and 4.0×10^4, respectively, at the end of two tests. These strain magnitudes are consistent with those previously documented in association with complete wetting of Hauzenberg granite [29] and other granitic rocks [27, 28]. This suggests that Herrnholz granite specimens behave in a similar manner to these previously tested granitic materials. Hydroscopic expansion observed in other rock types [27, 59, 60] is one (e.g., limestone and marbles) or several (e.g., sandstone) orders of magnitude greater than that observed in Herrnholz granite specimens, which could be associated with mineral composition, pore size distribution, and grain/micro-crack fabric.

4.5 Hygroscopic expansion and elastic weakening

Axial and radial expansion of the Herrnholz granite cylinder subjected to a stepwise increase in relative humidity was observed in both the 'staged' and 'continuous'

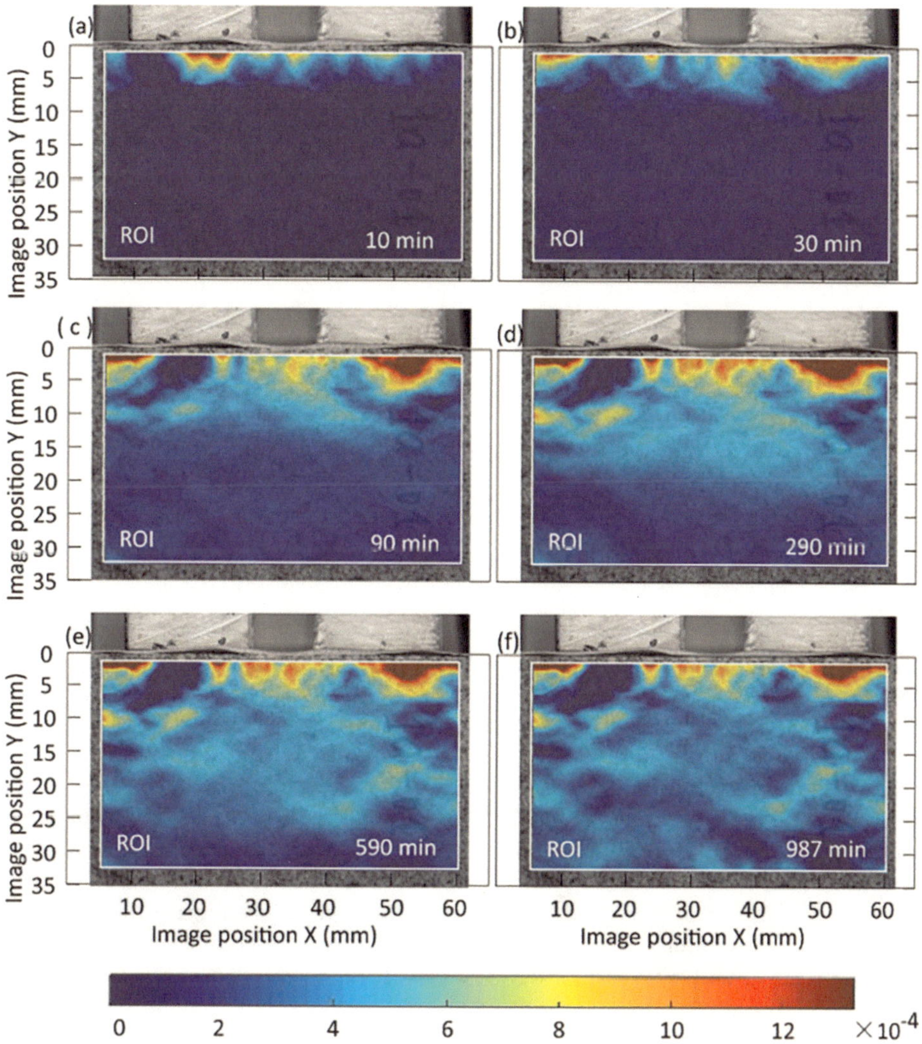

Figure 11.
Vertical strain evolution during water imbibition of a free-standing granite prism.

tests. Strains evolution over the 0.1 MPa hold periods (see test description in Section 3.4.2) provides information on the rock deformation in response to ambient humidity change. As is demonstrated by **Figure 12**, the axial strain during the staged test consistently increased with increasing humidity and typically stabilized after approximately 37 h at each humidity level. The response of radial strain is relatively smaller (max .2.0 $\times$ 10^4 vs. 3.0 $\times$ 10^4), and it continued to progress throughout each humidity interval. Calculated volumetric strain, assumed to be axial strain +2$\times$ radial strain, during the staged test consistently increased with increasing humidity, up to 3.0 $\times$ 10^4 at the end of the 90% humidity stage. The overall evolution of strains in the staged test compares favorably with that in the continuous test, with the maximum volumetric strain in the continuous test reaching a more-or-less stable peak of 8.0 $\times$ 10^4 at the end of 90% humidity level.

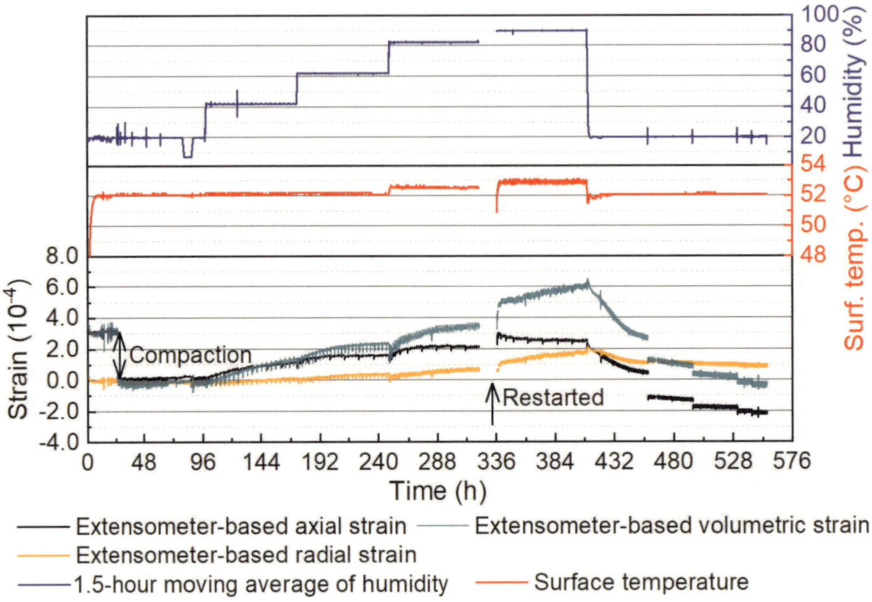

Figure 12.
Temporal evolution of strains (axial, radial, and volumetric) during hold periods across the 20–90% relative humidity range of the staged test.

Young's modulus and Poisson's ratio at varying humidity conditions were derived from axial stress–strain relationships during loading steps (see test description in Section 3.4.2) at each humidity level. Overall, Young's modulus in the staged and continuous tests decreased from ~ 43 to ~ 44 GPa, respectively, down to ~ 38 GPa in response to relative humidity increasing from 20% to 90% (see **Figure 13** for the staged test as an example). Returning humidity to 20% resulted in a recovery of Young's modulus, with a persistent 0.5 GPa increase in stiffness on completion of the staged test, and a similar decrease in stiffness on completion of the continuous test. Poisson's ratio (calculated from the staged test) increased from 0.12 to 0.26 across the humidity range, with a 60% recovery when the humidity level was reduced to 20% (**Figure 13**). Assuming an isotropic medium, bulk modulus from Young's modulus (E) and Poisson's ratio derived by $\frac{E}{3\times(1-2v)}$ increased from 18.8 GPa to 26 GPa, before decreasing to 22 GPa at the completion of the staged test (**Figure 13**).

4.6 Pore size effect on hydroscopic and hygroscopic properties

The deformation and elasticity variations of Herrnholz granite in response to the addition of liquid water or exposure to high humidity conditions have been commonly attributed to the physical process of water vapor adsorption [43, 49]. It exhibits different stages depending on the pore size within the absorbent media. In general, adsorption in micropores (pore diameter <2 nm based on Ref. [61]) is dominated by the interactions between the adsorbed fluid and pore wall due to the small pore size. Micropore filling is therefore continuous at low relative pressure [61], with a limiting amount adsorbed at the saturation pressure. In contrast, the size of meso- (2 nm < pore diameter <50 nm based on Ref. [61]) and macropores (pore diameter >50 nm

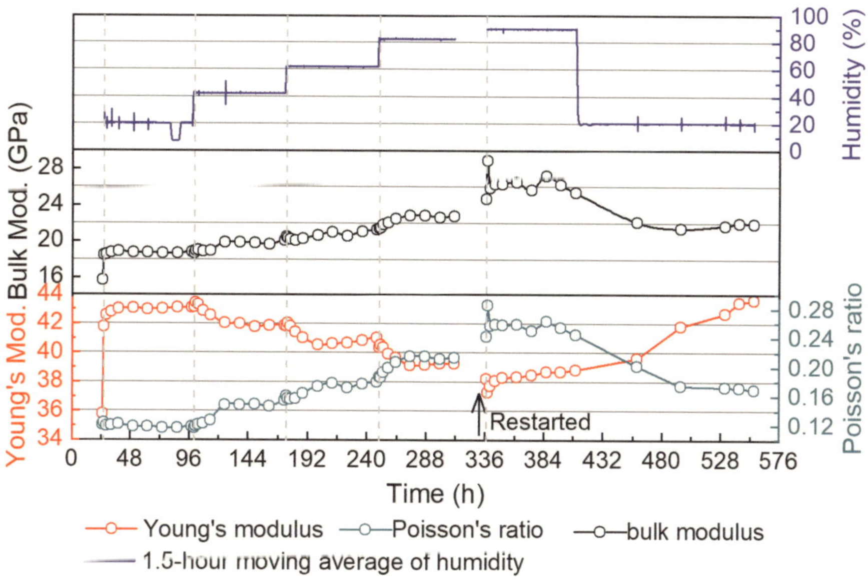

Figure 13.
Temporal evolution of Young's modulus, Poisson ratio, and bulk modulus (calculated from axial stress-strain measurement) for the staged test. The sharp change in Young's modulus immediately after the 1-h hold interval is indicated by dashed vertical lines.

based on Ref. [61]) is significantly larger than the effective range of adsorbent-adsorptive interaction (a few nm). Therefore, adsorption in meso- and macropores depends not only on the fluid-wall attraction, but also on the attractive interactions between the fluid molecules, leading to a sharp increase in the amount of adsorbate as relative pressure increases to a critical level, or even unrestricted monolayer–multilayer formation during adsorption. More precisely, Gor and Neimark [62] developed an adsorption model for mesoporous media, which describes the adsorption stress as a function of environmental conditions, pore size within the absorbent media, and adsorbate–-adsorbent interaction.

Combing the volumetric expansion ($\Delta\epsilon_{vol}$) and changing bulk modulus (K) with the assumption of a linear Hooke law, the adsorption stress ($\sigma_a = K\Delta\epsilon_{vol}$) developed from 'zero' to 25 GPa as relative humidity increases from 20% to 90%, agreeing well with modeled adsorption stress change for cylindrical pores with a characteristic diameter (i.e., 10 nm for Herrnholz granite, see Section 4.1). This model predicts a contrast of adsorption stress of a factor of ~ 2 between the 5 nm diameter and 10 nm diameter pores under the same environmental conditions for water vapor adsorption. We therefore expect larger adsorption-induced strains and elastic variations for intact rock containing smaller micropores or microfractures with smaller apertures.

5. Conclusions

Based on the material characterization results that involve microstructure distribution, acoustic wave velocity, elastic and brittle properties, and hydroscopic and hygroscopic properties, the following conclusions about the Herrnholz granite can be drawn:

1. fine-grained granite with grain sizes ranging from 0.03 to 1 mm (mean 0.23 mm, standard deviation 0.13 mm);

2. mesoporous media with a majority of 10 nm diameter mesopores providing 4.6 m^2/g specific surface areas;

3. isotropic wave propagation along three major splitting directions, with a mean velocity of 3982($\pm$64) m/s;

4. homogeneous material, as evidenced by a high degree of consistency in bulk density, porosity, stress–strain response, and fracture toughness across multiple samples;

5. detectable variations in strains and elastic properties caused by water vapor adsorption, which has not been widely recognized in rock mechanics or earth surface process research.

Acknowledgements

This work was supported by Swiss National Science Foundation (SNSF) R 'Equip "Long-term damage evolution in brittle rocks subject to controlled climatic conditions" (Project 170746) and "Physical constraints on natural and induced earthquakes using innovative lab-scale experiments: The LabQuake Machine" (Project 170766). The authors greatly thank Dr. Michael Plötze for the pore size distribution measurement at the Clay Lab of Institute for Geotechnical Engineering, ETH Zürich, and Markus Rast for the grain density measurement at the Rock Physics and Mechanics Lab, ETH Zürich.

Conflict of interest

The authors declare no conflict of interest.

Abbreviations

UCS	uniaxial compressive strength
SENB	single edge notch bending
DIC	digital image correlation
HQ	Herrnholz quarry
HP	Hauzenberg pluton
BET	Brunauer-Emmett-Teller
CI	crack initiation
CD	crack damage

Author details

Ying Li[1,2]* and Rui Wu[2]

1 Hebei University of Technology, Tianjin, China

2 ETH Zurich, Zürich, Switzerland

*Address all correspondence to: ying.li@erdw.ethz.ch

References

[1] Ramspott LD, Ballou LB, Carlson RC, Montan DN, Butkovich TR, Duncan JE, et al. Technical Concept for Test of Geologic Storage of Spent Reactor Fuel in the Climax Granite, Nevada Test Site. Livermore, CA (United States): California Univ; 1979. Lawrence Livermore Lab. UCID-18197

[2] Gnirk PF, McClain WC. An overview of geologic disposal of radioactive wastes. In: Bergman M, editor. Subsurface Space. Stockholm, Sweden: Pergamon; 1981. pp. 865-872

[3] Miller W, Alexander R, Chapman N, McKinley JC, Smellie JAT. Geological Disposal of Radioactive Wastes and Natural Analogues. Oxford, England: Elsevier; 2000

[4] Park ES, Jung YB, Song WK, Lee DH, Chung SK. Pilot study on the underground lined rock cavern for LNG storage. Engineering Geology. 2010; **116**(1):44-52

[5] Reid JC, Myers C, Carpenter RH. underground storage of refrigerated natural gas in granites of the Southeastern US. In: 47th Annual AAPG-SPE Eastern Section Joint Meeting, Pittsburgh, Pennsylvania

[6] Sibbitt WL, Dodson JG, Tester JW. Thermal conductivity of crystalline rocks associated with energy extraction from hot dry rock geothermal systems. Journal of Geophysical Research. 1979; **84**:1117-1124

[7] Kumari WGP, Ranjith PG, Perera MSA, Shao S, Chen BK, Lashin A, et al. Mechanical behaviour of Australian Strathbogie granite under in-situ stress and temperature conditions: An application to geothermal energy extraction. Geothermics. 2017;**65**:44-59

[8] Brunton GD, McClain WC. Geological Criteria for Radioactive Waste Repositories. Oak Ridge, Tenn. (USA): Union Carbide Corp; 1977. Office of Waste Isolation; Y/OWI/TM-47

[9] Tarkowski R. Underground hydrogen storage: Characteristics and prospects. Renewable and Sustainable Energy Reviews. 2019;**105**:86-94

[10] Matos CR, Carneiro JF, Silva PP. Overview of large-scale underground energy storage Technologies for Integration of renewable energies and criteria for reservoir identification. Journal of Energy Storage. 2019;**21**: 241-258

[11] Freire-Lista D, Fort R. Causes of scaling on bush-hammered heritage ashlars: A case study—Plaza Mayor of Madrid (Spain). Environmental Earth Sciences. 2016;**75**(10):1-12

[12] Freire-Lista DM, Fort R, Varas-Muriel MJ. Thermal stress-induced microcracking in building granite. Engineering Geology. 2016;**206**:83-93

[13] Freire-Lista D, Fort R. Exfoliation microcracks in building granite. Implications for anisotropy. Engineering Geology. 2017;**220**:85-93

[14] Sousa LMO. Petrophysical properties and durability of granites employed as building stone: A comprehensive evaluation. Bulletin of Engineering Geology and the Environment. 2014; **73**(2):569-588

[15] Sousa LMO, Suárez del Río LM, Calleja L, Ruiz de Argandoña VG, Rey AR. Influence of microfractures and porosity on the physico-mechanical properties and weathering of ornamental

granites. Engineering Geology. 2005; **77**(1):153-168

[16] Sousa LMO. The influence of the characteristics of quartz and mineral deterioration on the strength of granitic dimensional stones. Environment and Earth Science. 2013; **69**(4):1333-1346

[17] Dwivedi RD, Goel RK, Prasad VVR, Sinha A. Thermo-mechanical properties of Indian and other granites. International Journal of Rock Mechanics and Mining Sciences. 2008;**45**(3): 303-315

[18] David C, Menéndez B, Darot M. Influence of stress-induced and thermal cracking on physical properties and microstructure of La Peyratte granite. International Journal of Rock Mechanics and Mining Sciences. 1999;**36**(4): 433-448

[19] Heuze FE. High-temperature mechanical, physical and thermal properties of granitic rocks—A review. International Journal of Rock Mechanics and Mining Sciences & Geomechanics Abstracts. 1983;**20**(1):3-10

[20] Shao S, Wasantha PLP, Ranjith PG, Chen BK. Effect of cooling rate on the mechanical behavior of heated Strathbogie granite with different grain sizes. International Journal of Rock Mechanics and Mining Sciences. 2014; **70**:381-387

[21] Vázquez P, Shushakova V, Gómez-Heras M. Influence of mineralogy on granite decay induced by temperature increase: Experimental observations and stress simulation. Engineering Geology. 2015;**189**:58-67

[22] Vasconcelos G, Lourenço PB, Alves CAS, Pamplona J. Ultrasonic evaluation of the physical and mechanical properties of granites. Ultrasonics. 2008;**48**(5):453-466

[23] Cerrillo C, Jiménez A, Rufo M, Paniagua J, Pachón FT. New contributions to granite characterization by ultrasonic testing. Ultrasonics. 2014; **54**(1):156-167

[24] Korobiichuk I, Korobiichuk V, Hajek P, Kokes P, Juś A, Szewczyk R. Investigation of leznikovskiy granite by ultrasonic methods. Archives of Mining Sciences. 2018;**63**(1):75-82

[25] Vasconcelos G, Lourenço PB, Alves C, Pamplona J. Prediction of the Mechanical Properties of Granites by Ultrasonic Pulse Velocity and Schmidt Hammer Hardness. Luis, Missouri. 2007

[26] Ericson K. Geomorphological surfaces of different age and origin in granite landscapes: An evaluation of the Schmidt hammer test. Earth Surface Processes and Landforms. 2004;**29**(4): 495-509

[27] Hockman A, Kessler DW. Thermal and moisture expansion studies of some domestic granites. US Bureau Standards of Journal Research. 1950;**44**:395-410

[28] Swanson PL. Subcritical crack growth and other time-and environment-dependent behavior in crustal rocks. Journal of Geophysical Research: Solid Earth. 1984;**89**(B6): 4137-4152

[29] Schult A, Shi G. Hydration swelling of crystalline rocks. Geophysical Journal International. 1997;**131**(1):179-186

[30] Rivas T, Prieto B, Silva B. Influence of rift and bedding plane on the physico-mechanical properties of granitic rocks. Implications for the deterioration of

granitic monuments. Building and Environment. 2000;**35**(5):387-396

[31] Lindqvist JE, Åkesson U, Malaga K. Microstructure and functional properties of rock materials. Materials Characterization. 2007;**58**(11):1183-1188

[32] Nasseri MHB, Mohanty B. Fracture toughness anisotropy in granitic rocks. International Journal of Rock Mechanics and Mining Sciences. 2008;**45**(2): 167-193

[33] Klein T, Kiehm S, Siebel W, Shang C, Rohrmüller J, Dörr W, et al. Age and emplacement of late-Variscan granites of the western Bohemian Massif with main focus on the Hauzenberg granitoids (European Variscides, Germany). Lithos. 2008;**102**(3–4):478-507

[34] Siebel W, Shang C, Reitter E, Rohrmüller J, Breiter K. Two distinctive granite suites in the SW Bohemian Massif and their record of emplacement: Constraints from geochemistry and zircon 207Pb/206Pb chronology. Journal of Petrology. 2008;**49**(10):1853-1872

[35] Brunauer S, Emmett PH, Teller E. Adsorption of gases in multimolecular layers. Journal of the American Chemical Society. 1938;**60**(2):309-319

[36] Barrett EP, Joyner LG, Halenda PP. The determination of pore volume and area distributions in porous substances. I. Computations from nitrogen isotherms. Journal of the American Chemical Society. 1951;**73**(1):373-380

[37] Selvadurai PA, Wu R, Bianchi P, Niu Z, Michail S, Madonna C, et al. A methodology for reconstructing source properties of a conical piezoelectric actuator using Array-based methods. Journal of Nondestructive Evaluation. 2022;**41**(1):23

[38] Wu R, Selvadurai PA, Chen C, Moradian O. Revisiting piezoelectric sensor calibration methods using elastodynamic body waves. Journal of Nondestructive Evaluation. 2021;**40**(3):68

[39] Martiartu NK, Böhm C. TTomo: Straight ray tomography. Seismology and Wave Physics group at ETH Zurich; 2017. Available from: https://cos.ethz.ch/software/research/ttomo.html

[40] Li Y, Leith K, Moradian O, Loew S, Perras MA. A new laboratory to undertake climatically controlled static loading and constant strain tests: Design and preliminary results. In: 53rd US Rock Mechanics/Geomechanics Symposium. New York City, New York: OnePetro; 2019

[41] Diederichs MS, Martin CD. Measurement of spalling parameters from laboratory testing. In: Rock Mechanics and Environmental Engineering. Paper presented at European Rock Mechanics Symposium. Lausanne, Switzerland. 2010. pp. 323-326

[42] Martin CD. The Strength of Massive Lac du Bonnet Granite around Underground Openings. Canada: Ph.D University of Manitoba; 1993

[43] Li Y, Leith K, Perras MA, Loew S. Digital image correlation–based analysis of hygroscopic expansion in Herrnholz granite. International Journal of Rock Mechanics and Mining Sciences. 2021; **146**:104859

[44] Baud P, Zhu W, Wong T. Failure mode and weakening effect of water on sandstone. Journal of Geophysical Research: Solid Earth. 2000;**105**(B7): 16371-16389

[45] Chang C, Haimson B. Effect of fluid pressure on rock compressive failure in a

nearly impermeable crystalline rock: Implication on mechanism of borehole breakouts. Engineering Geology. 2007; **89**(3):230-242

[46] Rajabzadeh MA, Moosavinasab Z, Rakhshandehroo G. Effects of rock classes and porosity on the relation between uniaxial compressive strength and some rock properties for carbonate rocks. Rock Mechanics and Rock Engineering. 2012;**45**(1):113-122

[47] Nicolas A, Fortin J, Regnet J, Dimanov A, Guéguen Y. Brittle and semi-brittle behaviours of a carbonate rock: Influence of water and temperature. Geophysical Journal International. 2016;**206**(1):438-456

[48] Heap MJ, Farquharson JI, Kushnir ARL, Lavallée Y, Baud P, Gilg HA, et al. The influence of water on the strength of Neapolitan yellow tuff, the most widely used building stone in Naples (Italy). Bulletin of Volcanology. 2018;**80**(6):51

[49] Li Y, Leith K, Perras MA, Loew S. Effect of ambient humidity on the elasticity and deformation of unweathered granite. Journal of Geophysical Research [Solid Earth]. 2022;**127**(11):1-27

[50] Blaber J, Adair B, Antoniou A. Ncorr: Open-source 2D digital image correlation matlab software. Experimental Mechanics. 2015;**55**(6):1105-1122

[51] Labani MM, Rezaee R, Saeedi A, Al HA. Evaluation of pore size spectrum of gas shale reservoirs using low pressure nitrogen adsorption, gas expansion and mercury porosimetry: A case study from the Perth and Canning Basins, Western Australia. Journal of Petroleum Science and Engineering. 2013;**112**:7-16

[52] Brace W, Paulding B Jr, Scholz C. Dilatancy in the fracture of crystalline rocks. Journal of Geophysical Research. 1966;**71**(16):3939-3953

[53] Martin CD. Seventeenth Canadian geotechnical colloquium: The effect of cohesion loss and stress path on brittle rock strength. Canadian Geotechnical Journal. 1997;**34**(5):698-725

[54] Katz O, Reches Z. Microfracturing, damage, and failure of brittle granites: Microfracturing and failure of granites. Journal of Geophysical Research. 2004; **109**(B1)

[55] Nicksiar M, Martin CD. Evaluation of methods for determining crack initiation in compression tests on low-porosity rocks. Rock Mechanics and Rock Engineering. 2012;**45**(4):607-617

[56] Bieniawski ZT. Mechanism of brittle fracture of rock: Part I—Theory of the fracture process. International Journal of Rock Mechanics and Mining Sciences & Geomechanics Abstracts. 1967;**4**(4): 395-406

[57] Lajtai EZ. Brittle fracture in compression. International Journal of Fracture. 1974;**10**(4):525-536

[58] Moradian O, Wu R, Li Y, Leith K, Loew S. Acoustic emission and digital image correlation for damage evolution in brittle rocks under time-dependent tensile loading. In: IOP Conference Series: Earth and Environmental Science. Vol. 833. Turin, Italy: IOP Publishing; 2021. p. 012090

[59] Weiss T, Siegesmund S, Kirchner D, Sippel J. Insolation weathering and hygric dilatation: Two competitive factors in stone degradation. Environmental Geology. 2004;**46**(3): 402-413

[60] Koch A, Siegesmund S. The combined effect of moisture and temperature on the anomalous expansion behaviour of marble. Environmental Geology. 2004;**46**(3): 350-363

[61] Gregg S. Adsorption, Surface Area and Porosity. 2nd ed. London: Academic Press; 1982

[62] Gor GY, Neimark AV. Adsorption-induced deformation of mesoporous solids. Langmuir. 2010;**26**(16): 13021-13027

Section 2

Synthetic Minerals

Chapter 7

Production and Applications of Synthetic Quartz

Jun-Ichi Yoshimura

Abstract

Development of quartz industry in these 100 years, in the synthetic production and in application of quartz crystals, is surveyed briefly. Due to the discovery of piezoelectricity in 1880 by J. & P. Curie, the invention of a quartz resonator in 1921 by W. G. Cady, etc., it had been shown that quartz crystal can be used as a superior device material for electric wave telecommunication. After World War II ended, the civil use of quartz devices increased rapidly, in addition to the past military use. The study for large-scale industrial production of synthetic quartz was started first in the United States, and then spread out in other countries. In this development of synthetic quartz, the temperature-difference growth method that is used now was established. Large-sized quartz crystals of good quality (20–25 mm (X-direction) × 30–35 mm (Z-direction)) came to be obtained by around 1970, and now further large and good-quality crystals are produced. As another remarkable development in the quartz industry, the manufacture and spread of high precision quartz watch and clock should also be mentioned. Industry of synthetic quartz is progressing even now, and quartz devices are used in various places of our every-day life and of social activities.

Keywords: piezoelectric oscillation, telecommunication, quartz resonator, quartz crystal filter, synthetic quartz production, quartz watch, quartz clock

1. Introduction

For its good transparency and beautiful appearance, quartz has been used as a spiritual-powered stone in religious ceremonies and divination and has been valued as accessories and materials for accessories since the early days. It also has been used as an optical material for prisms. However, in the modern ages, the greatest application of quartz has been its use as an electric resonator owing to its highly precise and exact piezoelectric oscillation. Crystal material with such superior characteristics is not found besides quartz, although various crystalline materials have been developed for the electric resonator.

2. Invention of a quartz resonator and its subsequent developments

The quartz resonator stemmed from the discovery of piezoelectricity in a quartz crystal by Jacques and Pierre Curie (France) in 1880. Although the discovery had not been noticed directly, Paul Langevin (France) invented an ultrasonic-wave generator as an application of the piezoelectricity of quartz in 1917, finding that a strong inaudible sound wave is generated due to a responding piezoelectric flexural oscillation of a quartz plate when an alternating electric voltage is applied to it. The ultrasonic-wave generator was used for underwater communication and the search of submarines, etc., in World War I (1914–1918).

After that, in 1921, Walter G. Cady (USA) found that a quartz crystal shows a sharp resonance when changing the frequency of applied voltage to the natural frequency of it [1], and proposed to use such quartz resonators by incorporating them in an electric circuit of telegraphic communication (**Figure 1**), for the stabilization of the communication wave. The quartz oscillator was born thus. In 1923, George W. Pierce (USA) manufactured the so-called Pierce-type quartz resonator [2] by improving Cady's one [1], and this type of resonator came to be used widely in electric communication. The quartz resonator, which was invented by Cady, utilized the thickness longitudinal vibration of an *X*-cut plate (see **Figure 2(b)**). It was weak in the vibration intensity, and an oscillator that incorporated it had a fault that its operation often was not easy to start. The quartz resonator improved by Pierce utilized the thickness-shear vibration of a *Y*-cut plate and was large in vibration intensity and easy to start.

However, this *Y*-cut plate resonator also had a problem that the frequency temperature coefficient (hereafter, f-t coef.) was considerably large ($+100 \times 10^{-6}$/°C), so the resonant frequency shifted a little by a change in room temperature.

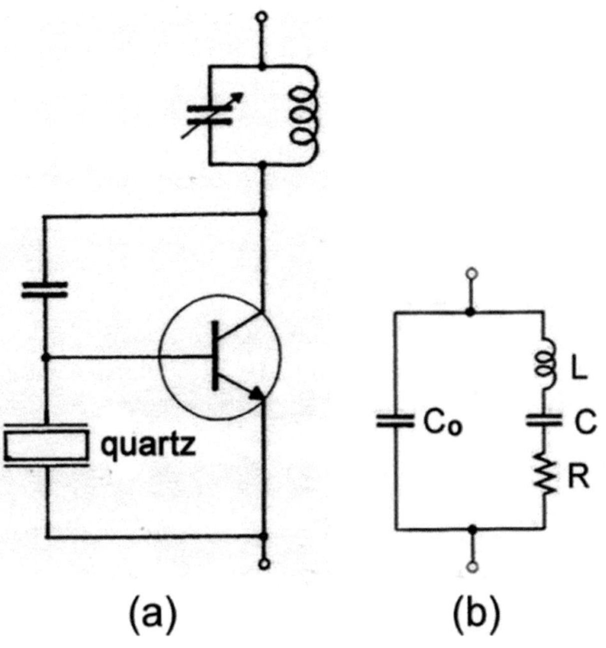

Figure 1.
(a) An example of the oscillating circuit where a quartz resonator is incorporated. (b) Equivalent electric circuit of the quartz resonator.

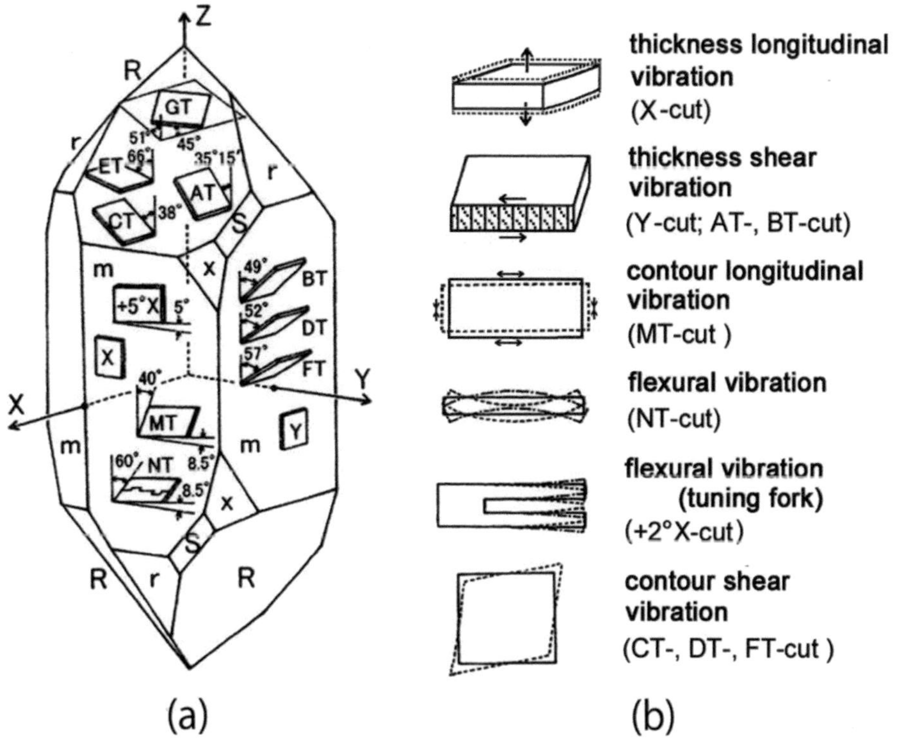

Figure 2.
(a) Cutting orientations of resonator plates of quartz crystal. X axis: [2-1-10], Y-axis [01-10], Z-axis: [0001] orientation. (b) Vibration modes of resonator plates.

Thus, search for the orientation of cut which has the zero f-t coef. characteristic began in the world. In this study, where many researchers joined, Issaku Koga (Japan) found in 1933 that the f-t coef. becomes zero in resonator plates cut out parallel to planes rotated around the X axis by 35° (more exactly, 35°15′) and − 49°, respectively, from the Y plane [3]. (Here, the zero f-t coef. means that the coefficient of the first-order term proportional to temperature is zero in the polynomial expression of the temperature dependence of the resonant frequency.) The two zero f-t coef. plates were named R_1 plate and R_2 plate, respectively, by Koga. Both of them are resonators of thickness-shear mode vibration. The f-t coef. of R_1 plate is less than 0.1×10^{-6}/°C at room temperature. One year later, in 1934, Bell Telephone Laboratories (USA) published the make of AT-cut and BT-cut plates as quartz resonators. However, they were the same as R_1 and R_2 plates, respectively, found by Koga. Nevertheless, the names of AT-cut and BT-cut are generally used in the world. Now, the AT-cut plate is representative of quartz resonator products.

3. Development of large-scale production of synthetic quartz resonator

3.1 Studies of synthetic quartz growth preceding large-scale development project

The zero f-t coef. quartz oscillator mentioned above was at once employed in World War II. The demand for quartz resonators for communication increased with the outbreak of the war, and the want for raw quartz for producing resonators came to be felt seriously in every country. Although natural quartz was produced in any region in the world, the source of good quality and large-sized raw quartz was limited to a few countries such as Brazil, etc. Thus, the dependence so far on natural quartz as the raw material for quartz products was reconsidered, and the production of synthetic quartz came to be demanded.

The experiment of the artificial synthesis of quartz was first made in 1845 by K. E. von Schanfhäutl (Germany). Gelatinous silica and water were sealed and heated up in an autoclave to obtain small quartz crystals of microscopic size. After that, several investigators made succeeding synthesis experiments. The experiment which directly led to the present growth technique of quartz was done by Giorgio Spezia (Italy) in 1905 [4, 5]. In his experiment, the growth was continued for 6 months in an autoclave under high temperature and high pressure, in which lasca quartz was charged as nutrient and a seed crystal was suspended in the Na_2SiO_3 solution. Through this experiment, he obtained a clear quartz crystal of 14 mm in length. In 1936, Richard Nacken (Germany) grew a quartz crystal of enough size to cut out a quartz resonator [6, 7], experimenting with a constant-temperature hydrothermal growth using amorphous silica, on the basis of Spezia's study.

3.2 Synthetic quartz production study in USA, starting after world war II

Although the necessity of raw quartz was understood, no project actually moved for the development of large-scale production of synthetic quartz, during World War II. After the war, demand for raw quartz was further increased with increasing civil use added to military use, so projects for large-scale production of synthetic quartz were planned. In the United States, a study for the industrial production of synthetic quartz was started in Brush Development Co., Ltd., Bell Telephone Laboratories, etc. [8], on the basis of Nacken's study. By these studies, synthetic quartz practically usable for telecommunication became producible in 1953, though at a research level.

As to the heating control of the autoclave, the constant-temperature method as used by Nacken and a temperature difference method were compared, and the latter method was concluded to be more suitable for mass production. In the latter temperature difference method, the autoclave is partitioned into two zones with a baffle between them, as known well (see **Figure 3(a)**); the one is the lower dissolving zone at a higher temperature (about 400°C) where nutrient lasca quartz is placed, and the other the upper growth zone at a lower temperature (about 350°C) where seed crystals are suspended. The temperature difference between the two zones was a few tens to several 10 degrees. As the working solution, a few to several weight % solutions of Na_2SiO_3, NaOH, etc., were used. As the seed crystal, a Y-directed slender bar-shaped crystal was concluded to be most suited, from various viewpoints; with this seed, quartz crystal grows fat in the perpendicular directions (X- and Z-directions) and not in the length direction (Y-direction) (see **Figure 3(b)**). This shape was favorable for producing many crystals in a tube-shaped autoclave. Furthermore, Y-bar quartz grown from Y-bar seeds was favorable for cutting out AT-cut plates efficiently without waste. Thus, the fundamentals of producing technology of synthetic quartz, which have succeeded to the present day, were established. The developing work for synthetic quartz production up to this time is well described in [8]. The developed manufacturing technology was then succeeded by Sawyer Research Products Co. Ltd. in USA. The company started as the first industrial maker of synthetic quartz crystal in the world, in 1956.

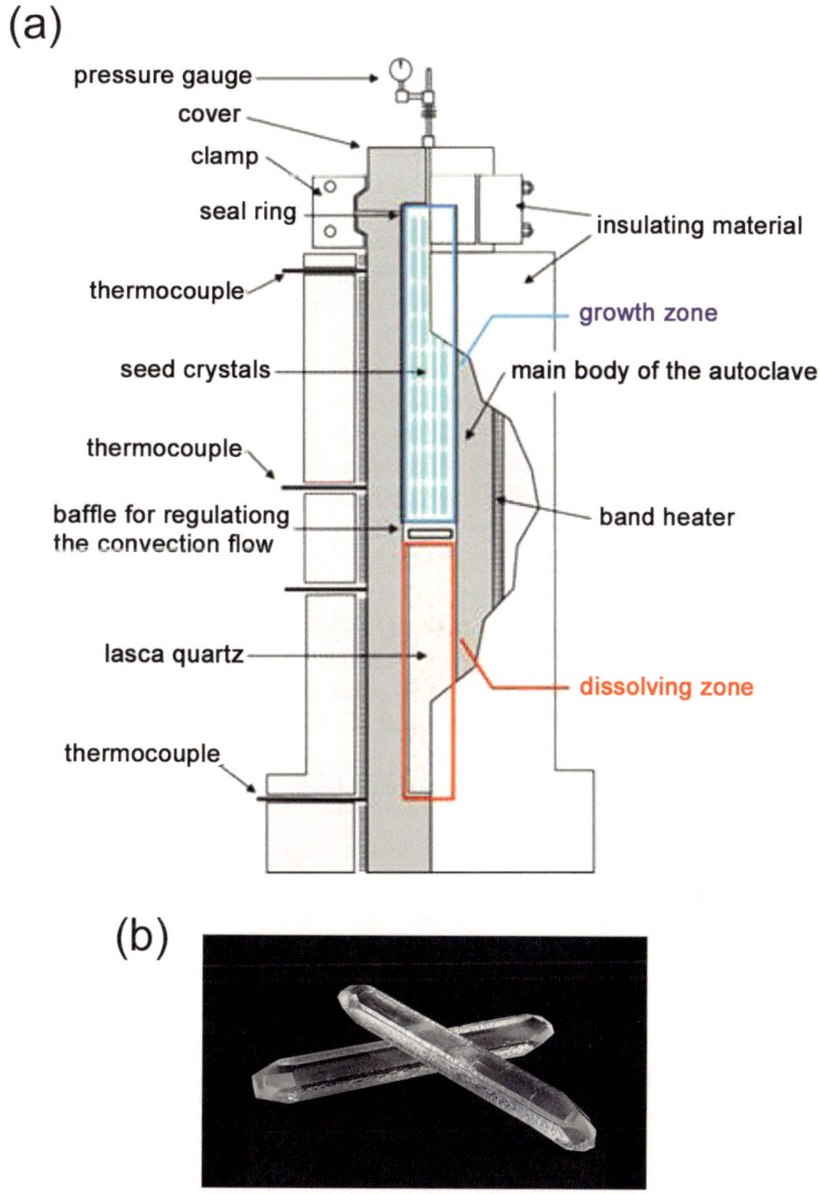

Figure 3.
(a) Autoclave for growing synthetic quartz crystals (of 650 mm in total inside diameter, and 14 m in total inside height). Figures cited from Reference [9]. (b) Grown Y-bar synthetic quartz crystals.

3.3 Synthetic quartz production in Japan

About the same time as above, the attempt to produce synthetic quartz also started in UK and France. In Japan, the research group of Minoru Kunitomi, Sadao Taki, and Junpei Asahara at Yamanashi University first succeeded in producing synthetic quartz in 1954 [10, 11]. The experiment was made by using an autoclave of 35 mm in inside diameter and 500 mm in inside depth. After that, Taki and Asahara transferred to Toyo Communication Equipment Co., Ltd. (hereafter, (Toyocom)) to participate in the technological development of industrial-scale production of synthetic quartz.

About the same time, several other companies such as Nihon Dempa Kogyo Co., Ltd. (hereafter, (NDK)), Kinsekisha Laboratory Corporation (hereafter, (Kinsekisha)), etc., also joined this business of industrial production of synthetic quartz. The technological development of quartz production by these Japanese companies reached an approximate completion in the early 1960s, but the effort to improve the technology for stably producing good quartz was further pursued.

In the hydrothermal growth of quartz a few to several weight % Na_2SiO_3 solution had mainly been used in a line from Spezia's experiment in 1905. However, in the development in (Toyocom) led by Taki, NaOH solution was employed. The gradient of the solubility curve to temperature is considerably lowered when the NaOH solution is used [12], compared with the case the Na_2SiO_3 solution is used. Accordingly, if the temperature difference set between the two zones in the autoclave is the same between the two cases above, the growth velocity with the NaOH solution is slower than that with the Na_2SiO_3 solution. However, in general consideration, better-quality crystals with the less defects can be obtained, corresponding to the slower growth. (Toyocom) chose to obtain better-quality quartz at the expense of some lowering of growth velocity. It is reported that obtained quartz quality was remarkably improved by this choice of growing solution in a report [10]. Another paper also reports that the clarity, or transparency, of NaOH-solution grown quartz was slightly higher than that of Na_2SiO_3 solution-grown quartz [13]. There would probably not yet be an exact comparative assessment of the qualities of the two kinds of quartz grown in NaOH and Na_2SiO_3 solutions. In Japan, while some makers other than (Toyocom) also employ NaOH solutions, other makers continue to use Na_2SiO_3 solutions.

3.4 Assessment of lattice defects of grown quartz crystals by X-ray topography, etc.

Around 1970, the development in Japan reached a goal line of producing apparently good quartz of enough size (20–25 mm (*X*-direction) × 30–35 mm (*Z*-direction)), and the concern of developers came to be also directed to the assessment of the quality of produced quartz. One of the to-be studied matters was the examination of lattice defects in the crystals, such as dislocations, and the related total assessment of the crystal quality. At that time, the assessment of lattice defects in silicon and other semiconductor crystals was still actively practiced using X-ray topography, etc. The study of lattice defects in quartz crystals entered such a current of study.

X-ray topographic studies disclosed that high-quality quartz crystals, which seemed to be usable as precision industrial materials, were grown in general [14–16] (**Figure 4**). Thin linear images in the X-ray topographs show dislocations. **Figure 4(a)** and **(b)**, respectively, show an example of a crystal with very few dislocations and a crystal containing many dislocations. Dislocations contained in a synthetic quartz crystal generally were not more than a few thousand/cm^2, if the crystal was grown with enough care, and they were in the range of several to several hundred/cm^2. Many dislocations were continued from ones in the seed crystal. Besides, many dislocations were newly generated along the seed boundary (see **Figure 4(b)**) in case the lattice spacing of the seed crystal has a large difference from that of the newly grown region. Few dislocations were generated on the way of growth.

Although each crystal of the present study is a single crystal as a whole, growth sectors $\pm Z$, $\pm X$, *s* are formed in its inside. Strong black or white linear contrasts along sector boundaries are caused by strains induced by the difference in lattice spacing between growth sectors, which arise from different impurity contents between the growth sectors.

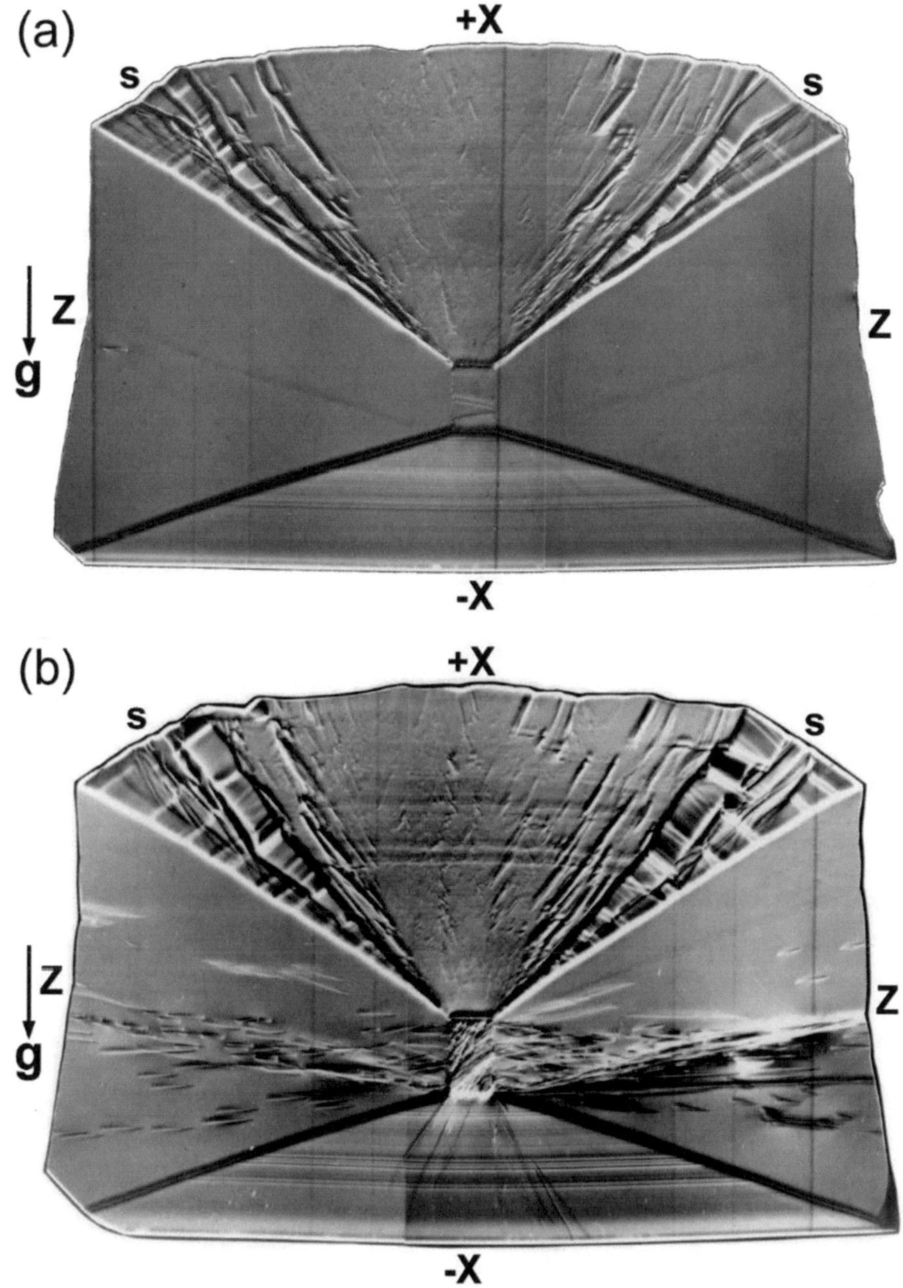

Figure 4.
X-ray diffraction topographs of Y-cut plates (0.3 mm thick) of Y-bar quartz crystals. Cu Kα radiation, -2020 reflection. g is the diffraction vector. The sizes of the crystals are 21-23 mm in the X-direction. Small rectangular regions at the centers of topographs respectively show the seed part of crystals.

Being different from the case of semiconductor crystals, the function of a quartz crystal as electrical device is not decisively influenced by one dislocation, but is influenced through a collective strain/stress effect from many dislocations. As the understanding at that time, dislocations as many as a tens/cm^2 were considered not to have a significant influence on the function of a quartz crystal. However, it was the knowledge up to 1990s. In the present day, after more than 30 years from that time, a study of the harmful influence of dislocations (screw type) is further progressed, and the influence is more seriously considered (see Section 4.3).

3.5 Progress in the crystal filter technology; the advent of MCF and SAW filters

Additionally, reference should be made to the progress in the technology of quartz crystal filters at that time. Now, a crystal filter and a quartz resonator are one of the major quartz products. Filters are used in electrical communication to take out or to cut only the selected bandwidth of frequencies from the received or transmitted electric waves. Initiatively used LC filters (inductor (L)-capacitor (C) electric circuit filters) were fabricated only from purely electric elements and were insufficient in the treatable frequency (10–40 kHz), frequency bandwidth, Q value characteristic, etc. For this condition of frequency filter, W.G. Cady suggested the use of quartz resonator as filter element in 1922 [17]. Quartz was expected to be used as superior filter material owing to its sharp resonance characteristic. In 1934 Walter P. Mason (USA) published his work on crystal filters [18], one of which (of "narrowband design") was widely used for the next 20 years. The one which is called the discrete type in literature is this type of filter (**Figure 5(a)**).

From the mid-1950s, demand arose for crystal filters usable for higher frequencies with a narrower bandwidth, and operating stably, from the extending use of filter elements in the military and civil-use apparatuses. Correspondingly, studies for the performance improvement of crystal filters came to be made actively. In such circumstances, in 1962, Yuzo Nakazawa (Japan) of (Toyocom) devised a new-type crystal filter, which is called afterward MCF (Monolithic Crystal Filter) [19, 20]. In this crystal filter, two sets of facing electrodes were installed on one quartz plate, and were acoustically connected to each other in the plate to realize the frequency filtering (**Figure 5(b)**).

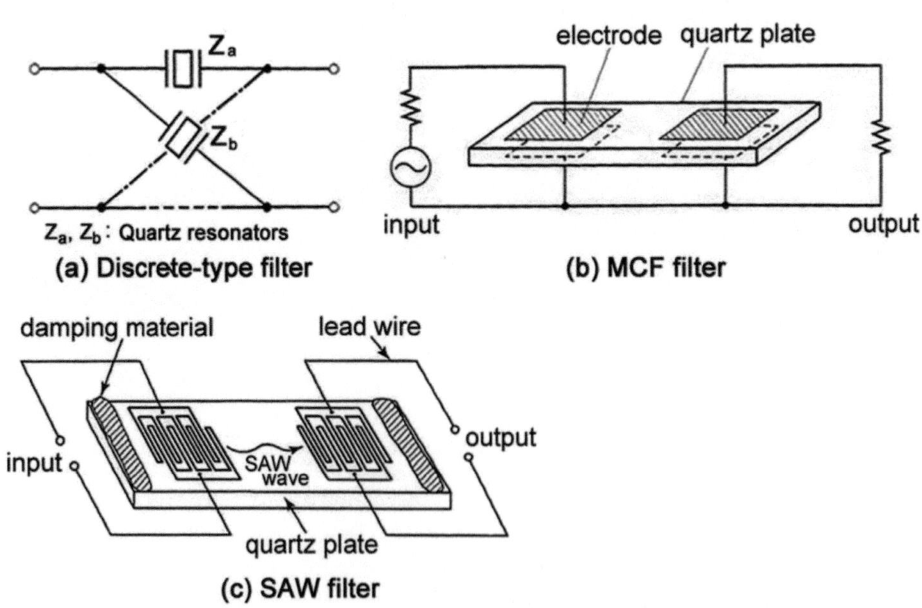

Figure 5.
Crystal filters as a quartz device.

The advent of this small and light crystal filter with a superior filtering characteristics attracted a great deal of public attention, and its use spread quickly to the world.

(MCF) was first employed as filters in ultra-high frequency radio communication. With the succeeding advances in radio communication, the central frequency in crystal filters was demanded to become further higher, from 10.7 MHz at the beginning. For the treated frequency to be made higher, the quartz plate needs to be made thinner. However, thinning processing had a technological limit then, and 70 MHz was a limit of realizable high frequency. For this difficulty, a crystal filter utilizing SAW wave (surface acoustic wave) was considered as a new (MCF) which oversteps the frequency limit (**Figure 5(c)**); the (SAW) wave excited in the surface of the quartz plate was utilized instead of oscillation in the bulk of the quartz plate. Though the subsequent history of the development of such (SAW) filters cannot exactly be followed here, it was in 1984 that (Toyocom) put (SAW) filters on the market, according to a report published by (Toyocom) [20]. In the present day (SAW) filters are used in almost all cellular phones. The present treatable frequency limit reaches several GHz. Additionally, while quartz (SAW) filters are used for narrow-bandwidth communication and with good temperature stability, (SAW) filters made of $LiTaO_3$ and $LiNbO_3$ crystals are more preferably used with wider bandwidths.

4. The present state of production of synthetic quartz and quartz devices

4.1 The present state of synthetic quartz production industry

Compared with the state about 50 years ago (about 1970) when the technology of synthetic quartz production was established, the present state of the industry and technology of synthetic quartz had been changed not a little, although the fundamentals of the technology was unchanged. The present world production of synthetic quartz is reported to be about 3200 t/year, according to a report in 2021. Major producing countries are China, Russia, and Japan. As for producing companies, (Sawyer Research Products) Inc. (*now* (Sawyer Technical Materials) LLC.) in USA is still in a healthy condition and is doing good work. In Japan, (Toyocom) Co., Ltd. was affiliated with (Seiko Epson) Corp., but the production of quartz devices is still actively continued by subsidiary companies (Epson-Toyocom) Corp., etc. (Kinsekisha) Laboratory Corp. also was affiliated with (Kyocera) Corp., and the production of quartz devices is continued as a business of (Kyocera) Corp. (NDK) Co., Ltd. continues the production of synthetic quartz and quartz devices as before, without merging with other companies.

4.2 Technological changes in the production of synthetic quartz and quartz devices

In the technology of producing synthetic quartz, the autoclave for growing quartz had been increased in size year by year, in order to cut down production costs. For example, about 1960, its typical size was 2 m in depth and 120 mm in inside diameter, whereas about 25 years later, in 1984, its depth was increased to 14 m with the inside diameter of 650 mm [9, 10]. The size and shape of seed crystal, and accordingly those of grown crystal, were also changed considerably. The size of the conventional Y-bar seed (about 3 mm in the X-direction, and a little less than 2 mm in the Z-direction) was greatly extended in the X-direction to lengths from 70 mm to more than 100 mm, and such seeds are called Z-plate seed. Using seed crystals like this, quartz crystals are grown to much larger than before (see **Figure 6**), up to volumes from several times

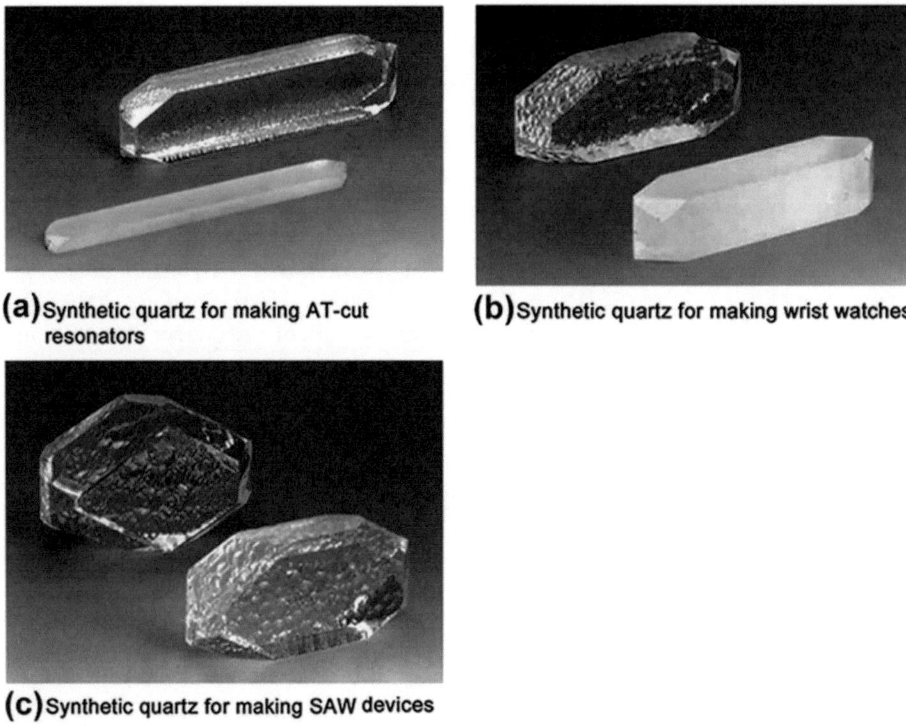

Figure 6.
Synthetic quartz crystals of various shapes and sizes, grown so as to match the purposes of use, in the present-day industrial production.

to dozens of times larger than that of former *Y*-bar crystals (**Figure 3(b)**); the weight of one crystal becomes 1–6 kg, while that of former *Y*-bar crystal is estimated to have been about 0.24 kg. One operating time of the autoclave is 2 to 6 months, while it was about 45 days (one and a half months) in the case of former *Y*-bar quartz. Though equally classified into the same category as *Z*-plate quartz, crystals are grown to separate sizes and shapes most suitable to the uses after the growth, for the manufacturing of *AT*-cut resonators, quartz watches, SAW devices, etc (see **Figure 6**).

As for manufactured quartz devices, the downsizing and thinning of quarts resonators is progressed, so that the devices of less than 1 mm in the substrate size are produced, and the thickness of the substrate is thinned down to 10 μm (the thinning is required for the use in high frequency). Many resonators and other devices of quartz are fabricated by lithographic processing.

4.3 The frame seed method for inhibiting the harmful occurrence of etch channels

Regarding dislocations in quartz crystals, it came to be known that, in the etching process in the blank-crystal processing, dislocations cause etch channels and etch pits in crystals to have a bad influence on the characteristic of resonators. In Japan, according to the Japanese Industrial Standard (JIS), the quality of quartz crystals is graded into six ranks based on the number 2–300 of etch channels contained in an area of 1 cm^2 in AT-cut plates [9]. Such etch channels and etch pits are caused by only screw dislocation, and edge dislocations do not cause them. Most of those screw dislocations succeed from the ones in the seed crystal. Thus, the choice of seed crystals is important. However, obtaining good seed crystals from natural quartz has become difficult since the production of large, high-quality natural quartz has become scarce. Instead, seed crystals also must be prepared from synthetic crystals. One method was developed for this seed crystal preparation to grow preferentially only the +X growth region, where the generation of a screw dislocation is difficult, using seed crystals of a special shape. Such +*X* growth region is used to cut out seed crystals for growing quartz crystals of ultimate products. This way of growing quartz crystals containing no- or few dislocations is called the frame seed method [9, 21].

5. The present state of extending use of quartz crystals

5.1 General view

Finally, we briefly remark on the present state of the use of quartz crystals. Now the whole world enters into a highly information-oriented society, where radio waves carrying various information travel around us everywhere and every day. A telecast of a sporting event carried out in a spot site on the earth is delivered worldwide via satellite broadcast, so that several billions of people enjoy watching the sporting event in real-time. On the surface of the ground, radio waves of cellular phones, wireless LAN, SNS, car navigation, etc., travels *to and from* busily. On the sea ship, radio waves travel about, and in the sky plane, radio waves travel. In such all radio and wired communications, quartz devices work efficiently for the exact and stable transmission, reception, and processing of electric waves. According to a recent guidance article, several tens of quartz devices are installed in one car, and they become several hundred in a luxury-model car.

5.2 Use in other than the communication field; quartz watch and clock

We next mention the use of quartz other than in the communication field. The largest field of such use of quartz would certainly be the use for quartz watch and quartz clock. The quartz crystal oscillates very regularly at a constant frequency by its piezoelectricity, when an AC voltage is applied to it. The fabrication of quartz clocks had early been made utilizing this exact regular oscillation. It is reported that a quartz clock was first constructed using a quartz resonator by Warren Marrison and J.W. Horton of Bell Telephone Laboratories (USA) in 1927 [22]. However, the clock was as large as a wardrobe because of the many vacuum tubes installed in the clock body, and its use was limited to special institutions such as (NBS) (National Bureau of Standards). Also in Japan, a high-precision quartz clock of a large size had been constructed similarly and installed in a special institute; it was used for a special purpose, not like being used by general people.

(Suwa Seikosha) Co., Ltd. (*now* (Seiko Epson) Corp.) was challenged to produce such quartz clocks of the size of a wristwatch, and succeeded in producing them after the development effort for ten years since 1958 [23]. The products were put on sale as "quartz watch" from 1969. When the error of general high-precision mechanical watches was several tens s/day, the error of this quartz watch was ±0.2 s/day, and ± 5 s/month. A key to the successful development was the employment of a tuning-folk type resonator (of flexural-vibration mode) (see **Figure 2(b)**), with which the resonance frequency could be greatly decreased, compared with the case when AT-cut resonators were used. (Now the frequency of 32,768 vibrations/s becomes the standard.) (Although, in the firstly presented quartz watch in 1969, a *Y*-shaped resonator was used, it was soon after replaced by the tuning-folk type resonator.) Tuning-folk type resonators (about 100 μm thick) are produced by lithographic processing.

In the 1970s (Suwa Seikosha) released the process patent for quartz watch, so many makers entered the business of producing quartz watch. As a result of expanding production, the price down of quartz watches advanced. The price of a quartz watch was 450,000 Japanese yen in the opening sale in 1969, but it went down rapidly to nearly 1000 yen at the lowest price, in 10 years or so. Now, not only quartz watch, but wall clocks, table clocks, alarm clocks, etc., all are produced as a quartz clock. With this development of quartz watch/clock industry, the conventional industry of mechanical clock/watch, which had developed in Europe and other regions in the world since the 16th century, came to decline. Although it is never a desirable trend of the times, it would be an affirmable progress in society that everyone in the world can wear a high-precision wristwatch at a little expense. Recently, the value of classic-style wrist watches is reconsidered as an accessory, so the manufacture of such mechanical watches is reported to be revived in part.

5.3 Use in other than the communication field; quartz crystal microbalance (QCM)

Whereas quartz watches above utilized the accuracy of piezoelectric vibration of quartz crystal, an application of a small change in the piezoelectric vibration frequency is Quartz Crystal Microbalance (QCM). Even when a trace of material adheres or leaves from the electrode, the correspondingly occurring microgravity change can be detected through a frequency change in the resonator. The basic theory of this measurement method has been known since 1959 [24]. However, it was about 30 years ago (1990s) that the method began to attract attention as a high-sensitivity technique for microgravimetry [25]. According to the theory, if the resonant frequency of the quartz plate (AT-cut) is 10 MHz, an increase of 1 Hz in the resonant frequency corresponds to a decrease of about 4 ng/cm^2 in the mass. This sensitivity is evaluated to be higher than that of a conventional microbalance. Besides, the sensitive frequency change is also utilized in film-thickness meters, pressure sensors, etc.

6. Conclusion

As seen, quartz devices are used in various places of our everyday life and social activities now. Before, silicon semiconductor chips were compared to “rice of industry.” That simile lives even now. Following it, quartz crystals are now compared to “salt of industry.” The use of quartz crystals would extend further along with the development of a high information-oriented society.

DOI: http://dx.doi.org/10.5772/intechopen.111593

Author details

Jun-Ichi Yoshimura
Photon Factory, Institute of Material Structure Science, High Energy Accelerator Research Organization (Former researcher), Tsukuba, Japan

*Address all correspondence to: j-yoshimura@voice.ocn.ne.jp

References

[1] Cady WG. The piezoelectric resonator. Physical Review. 1921;**A17**:531-533

[2] Pierce GW. Piezoelectric crystal resonators and crystal oscillators applied to the precision calibration of wave meters. Proceedings of the American Academy of Arts and Sciences. 1923;**59**:81-106

[3] Koga I. Piezoelectricity and High Frequency. Tokyo: Ohmsha; 1938

[4] Spezia G. Contribuzioni di geologia chimica. Atti della Accademia delle Scienze di Torino. 1905;**40**:254-262

[5] Spezia G. Contribuzioni sperimentali alla cristallogenesis del quartz. Atti della Accademia delle Scienze di Torino. 1906;**41**:158-165

[6] Nacken R. Die hydrothermalen Mineralsynthese als Grundlage zur Züchtung von Quartzkristallen. Chmiker-Ztg. 1950;**74**:745-749

[7] Nacken R. Die wirtschaftlich Bedeutung der hydrothermalen Forschung. Technisch Mitteilunger Heft. 1953;**4**(46):87-94

[8] Ballman AA, Laudise RA. Hydrothermal growth. In: Gilman JJ, editor. The Art and Science of Growing Crystals. New York: John Wiley & Sons. Inc; 1963. pp. 231-251

[9] Available from: https://www.ndk.com/jp/news/article/post_16.html

[10] Available from: https://www5.epsondevice.com/ja/information/technical_info/qmems/story2_4.html

[11] Taki S, Kunitomi M, Asahara J, Ozawa T. Hydrothermal growth of quartz. Kogyo Kagaku Zasshi. 1956;**59**:1337-1340

[12] Taki S, Yamada K. Solubility of quartz to the solutions of NaOH-Na_2CO_3-NaCl systems under high temperature and high pressure. Reports of the Faculty of Liberal Arts Yamanashi University. 1961;**12**:146-150

[13] Ohara G. Growth of quartz and its characteristics as a resonator. Yogyo-Kyokai-Shi. 1969;**77**:111-123

[14] Yoshimura J, Kohra K. Studies on growth defects in synthetic quartz by X-ray topography. Journal of Crystal Growth. 1976;**33**:311-323

[15] Yoshimura J, Miyazaki T, Wada T, Kohra K, Hosaka M, Ogawa T, et al. Measurement of local variations in spacing and orientation of lattice plane of synthetic quartz. Journal of Crystal Growth. 1979;**46**:691-700

[16] Yoshimura J. Growth-caused defects in synthetic quartz observed by X-ray topography/growth-induced defects in natural quartz observed by X-ray topography. In: Crystal Growth Handbook, Ed. Editing Committee of "Crystal Growth Handbook" in Japanese Association of Crystal Growth. Tokyo: Kyoritsu Shuppan; 1995. pp. 179-183

[17] Cady WG. The piezo-electric resonator. Proceedings of the Institute of Radio Engineers. 1922;**10**:83-114

[18] Mason WP. Electrical wave filters employing quartz crystals as elements. Bell System Technical Journal. 1934;**13**:405-452

[19] Nakazawa Y. High frequency crystal electro-mechanical filters. In:

Proceedings of 16th Annual Frequency Control Symposium. New York: Institute of Electrical and Electronics Engineers (IEEE); 1962. pp. 373-390

[20] Available from: https://www5.epsondevice.com/ja/information/technical_info/qmems/story2_5.html

[21] Mikawa Y, Hatanaka M, Banno Y. New technique to decrease dislocations in synthetic quartz crystal. In: Proc. 1999 Joint Meeting of EFTF-IEEE IFCS. 1999. pp. 773-776

[22] Marrison WA, Horton JW. Precision determination of frequency. In: Proceedings of the Institute of Radio Engineers. Vol. 16. New York: Institute of Radio Engineers; 1927. pp. 137-154

[23] Available from: https://ja.wikipedia.org/wiki/quartzwatch

[24] Sauerbrey G. The use of quartz oscillator for weighing thin layers and for microweighing. Zeitschrift für Physik. 1959;**155**:206-222

[25] Seo M. Quartz crystal microbarance. Hyomen Gijutsu. 1994;**45**:1003-1008

Chapter 8

Dislocation-Point Defects-Induced by X-Irradiation Interaction in Alkali Halide Crystals

Yohichi Kohzuki

Abstract

Strain-rate cycling tests under superposition of ultrasonic oscillation were carried out at 77–293 K for two kinds of samples: non-irradiated and X-ray-irradiated KBr single crystals. Point defects induced by X-ray irradiation have weak interaction with dislocation and act as obstacles to dislocation motion. Assuming that the defect is tetragonal, the interaction energy for the break-away of a dislocation from the defect has been obtained by fitting the Barnett model to experimental results. Then, the value of interaction energy was determined to be 0.81 eV for the crystal. This result is compared with it in other crystals (NaCl, NaBr, and KCl) by the X-irradiation.

Keywords: dislocation, X-ray irradiation, point defects, activation energy, Barnett model

1. Introduction

It is well known that alkali halide crystals are hardened by X-ray irradiation [1–4]. The defect due to the radiation is electron-centre or hole-centre such as F-centre, V_2- or V_3-centre. Although the yield stress becomes large with the irradiation dose [5], the interaction between dislocation and the defect is not clearly established yet.

In this chapter, the dislocation-radiation defects interaction in KBr single crystals is described by analysing the data obtained by the combination method of strain-rate cycling tests and ultrasonic oscillation. Useful information on the interaction between a mobile dislocation and additive ions has been reported so far for alkali halide crystals during plastic deformation by the method [6–8], which can separate the effective stress due to the additions from that due to dislocation cuttings.

2. Experimental procedure

2.1 Preparation of samples

The samples were prepared by cleaving out of KBr single crystalline ingot, which was grown from the melt of superfine reagent powder by the Kyropoulos method in air, to the size of $5 \times 5 \times 15$ mm^3. The samples were annealed at 973 K for 20 h and were gradually cooled to room temperature at the rate of 40 K/h in order to reduce dislocation density as much as possible. The samples were exposed to X-ray (W-target, 30 kV, 20 mA) for 3 h on each of the pair wide surfaces at room temperature by Shimadzu XD-610. Namely, the total exposure time is 6 h.

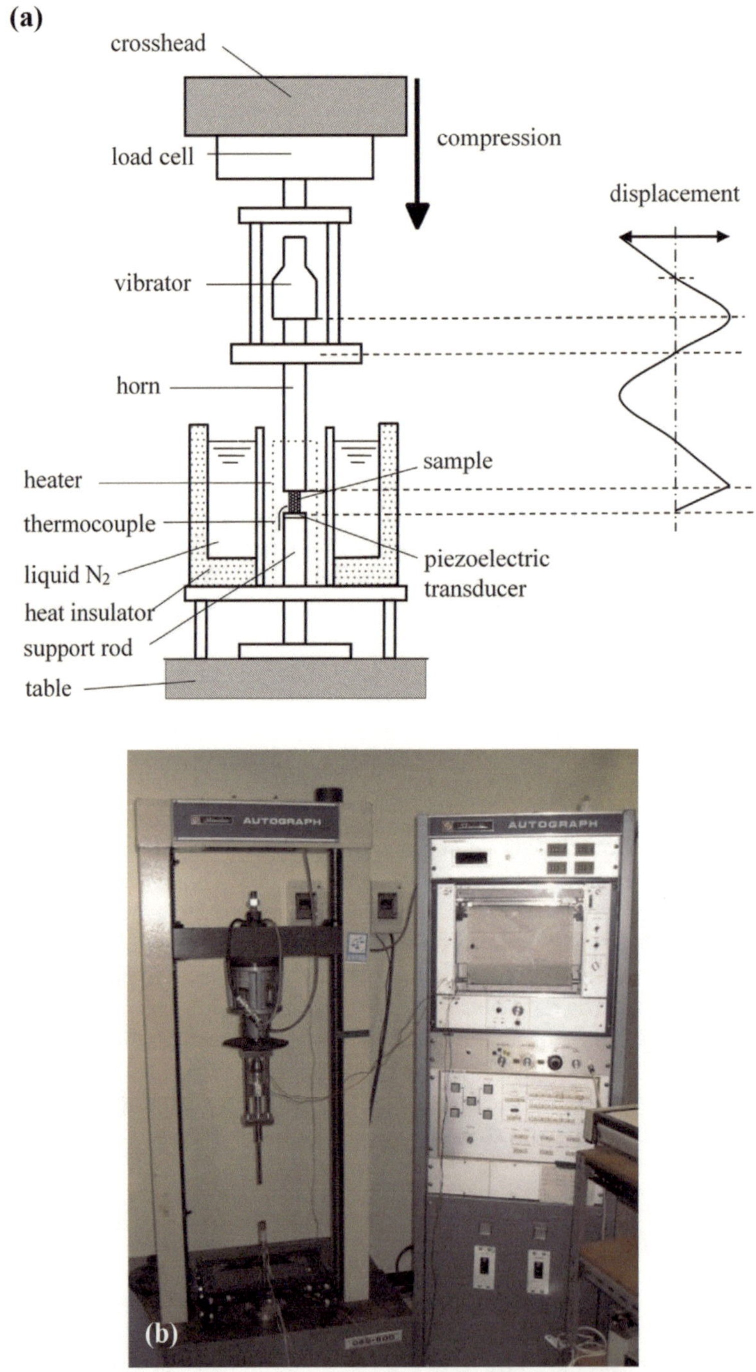

Figure 1.
Experimental apparatus: (a) schematic illustration of the apparatus, and (b) main testing machine.

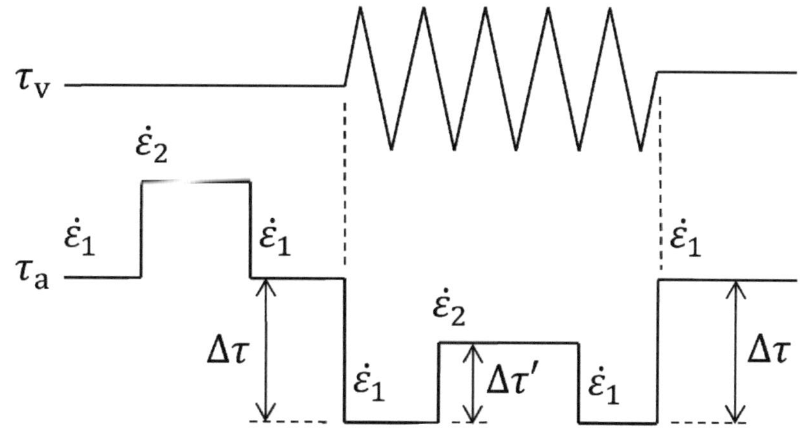

Figure 2.
Applied shear stress (τ_a) variation by the strain-rate cycling tests between the two strain rates, $\dot{\varepsilon}_1$ ($1.1 \times 10^{-5}\,s^{-1}$) and $\dot{\varepsilon}_2$ ($5.5 \times 10^{-5}\,s^{-1}$), off or on ultrasonic oscillatory shear stress (τ_v) due to the oscillation (20 kHz) during compression.

2.2 Strain-rate cycling tests under superposition of ultrasonic oscillation

The experimental apparatus is schematically illustrated in **Figure 1a**, where the resonator composed of a vibrator and a horn with a resonant frequency of 20 kHz was attached to a testing machine (Shimadzu DSS-500). **Figure 1b** shows the main testing machine. The samples, which were fixed on a piezoelectric transducer, were compressed along <100> direction of the longest axis of a crystal and the ultrasonic oscillatory stress was intermittently superimposed by the resonator in the same direction as the compression. The amplitude of the oscillatory stress τ_v was monitored by the output voltage from the piezoelectric transducer. The strain of the specimen seems to be homogeneous, since the wavelength (152 mm [9]) for KBr is approximately 10 times as long as the length of the sample.

The strain-rate cycling tests associated with ultrasonic oscillation are illustrated in **Figure 2**. Superposition of oscillatory stress τ_v causes a stress change ($\Delta\tau$) during plastic deformation. When strain-rate cycling between the strain rates of $\dot{\varepsilon}_1$ ($1.1 \times 10^{-5}\,s^{-1}$) and $\dot{\varepsilon}_2$ ($5.5 \times 10^{-5}\,s^{-1}$) was conducted, keeping the stress amplitude constant at 77 K to room temperature, the stress increment due to the strain-rate cycling is $\Delta\tau'$. The strain-rate sensitivity λ of flow stress was derived from the $\Delta\tau'$ value (that is to say, $\lambda = \Delta\tau'/\Delta\ln\dot{\varepsilon} = \Delta\tau'/\ln(\dot{\varepsilon}_2/\dot{\varepsilon}_1)$).

3. Results and discussion

3.1 X-ray-irradiated crystal

The sample was exposed to X-ray for 3 h on each of the pair wide surfaces at room temperature (i.e., total exposure time is 6 h) and was cleaved in four thin crystal plates (a)∼(d) at regular intervals (a thickness of about 1 mm) as illustrated in **Figure 3**. The concentration distribution of the F-centre (trapped electron) in the X-ray-irradiated crystal is shown for each plate in **Figure 3**. The abscissa "Distance from edge" of the figure represents the distance from the irradiated crystal surface of plate (a). The F-centre concentration, which was estimated from the Smakula formula [10], tends

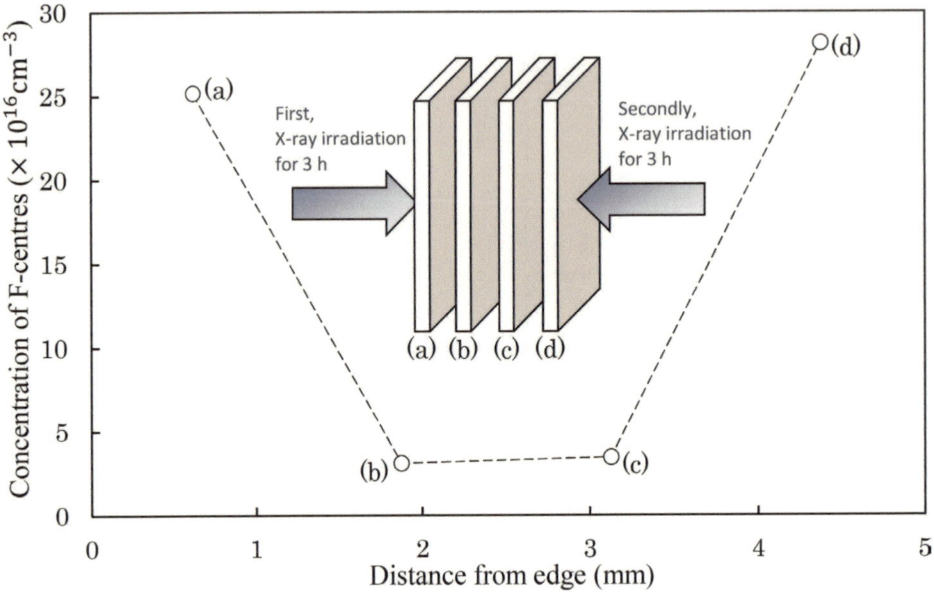

Figure 3.
Concentration distribution of F-centres in X-ray-irradiated KBr single crystal.

to decrease in the deep inside as against the surface of the sample. The average concentration of F-centres is 13×10^{16} cm^{-3} for the irradiated KBr crystal. However, F-centres or vacancies are so weak to interact with dislocation that they do not act as obstacles to dislocation motion, as described by Zakrevskii and Shul'diner [11]. This would be due to the isotropic defects around them in the crystal. Sirdeshmukh et al. also suggested that the radiation hardening is caused by the role of radiation-induced defects other than F-centres [12].

Figure 4 shows the absorption spectrum of the X-ray-irradiated KBr crystal. The colour centre has been reported for alkali halide [13, 14]. As can be seen in **Figure 4**,

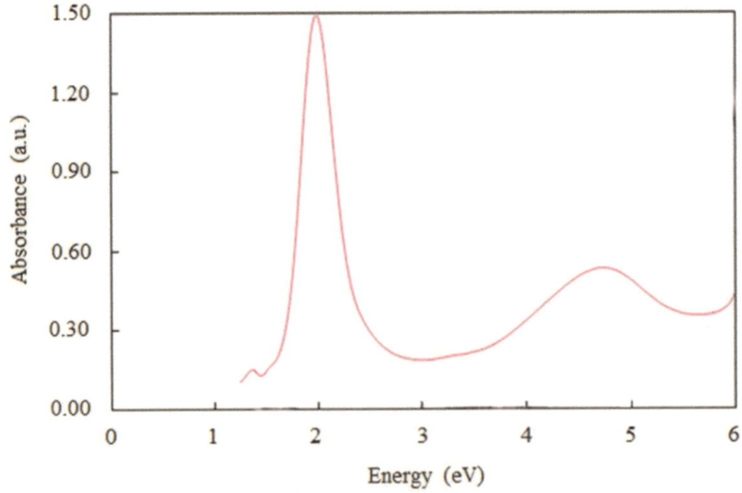

Figure 4.
Absorption spectrum of X-ray-irradiated KBr single crystal.

the peak of F-band (2.0 eV) is seen, and the peak of 4.7 eV is attributed to V_2-centre in the spectrum of X-irradiated KBr crystal. Since a hole-centre such as V_2-centre is considered to have not isotropic but tetragonal distortions, the irradiation-induced defects (V_2-centres) act as stronger obstacles to dislocation motion in comparison with F-centres or vacancies in the sample (the X-irradiated KBr crystal).

The radiation effect on stress–strain curve is shown in **Figure 5** for KBr crystal at room temperature. The curves (a) and (b) in the figure represent nonirradiated and X-ray-irradiated samples, respectively. When the crystal is exposed to the X-irradiation, flow stress increases at a given strain and the radiation hardens the crystal. This is because X-ray-induced point defects obstruct dislocation motion.

3.2 Effective stress (τ_p) due to X-ray-induced defects

The variations of $\Delta\tau$ and λ with shear strain ε are shown in **Figure 6a** and **b** for the non-irradiated KBr single crystal at 193 K. The oscillatory stress τ_v becomes large with output voltage, which is denoted in the legend of these figures, from the piezoelectric transducer. $\Delta\tau$ increases with the amplitude of τ_v and is almost constant independently of strain. λ tends to decrease with the stress amplitude, and the variation of it with τ_v is small at high amplitude. λ also increases with strain. This is due to the increase of the forest dislocation density with strain, since λ is proportional to the inverse of average length of dislocation segments.

The values of $\Delta\tau$ and λ at shear strains of 8, 12, 16, and 20% in **Figure 6a** and **b** are plotted in **Figure 7** as the relative curve of λ versus $\Delta\tau$ at a given strain. Only one bending point is on each curve.

The relation between λ and $\Delta\tau$ obtained by the above-mentioned method is shown in **Figure 8a** for the X-irradiated KBr single crystal at 183 K. **Figure 8b** corresponds to it for the irradiated KBr crystal at 203 K. The λ varies with $\Delta\tau$ like stair shape: the first

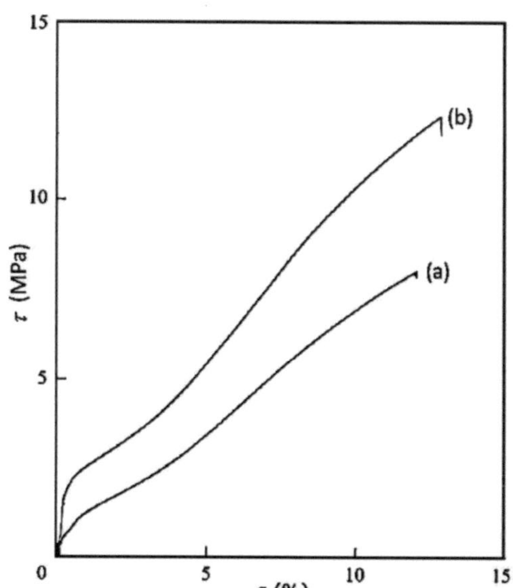

Figure 5.
Stress versus strain curve for (a) non-irradiated and (b) X-ray-irradiated KBr single crystals at room temperature.

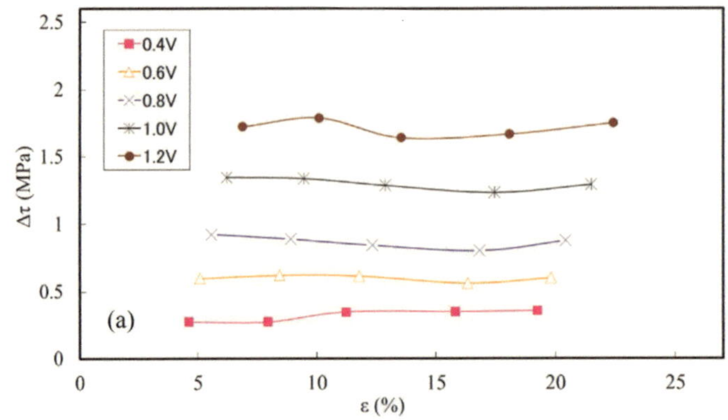

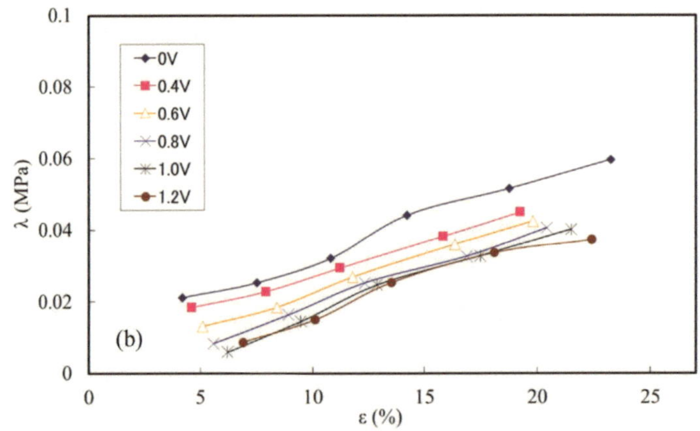

Figure 6.
Strain versus (a) stress decrement ($\Delta\tau$) due to superposition of oscillation and (b) strain-rate sensitivity (λ) of flow stress given by strain-rate cycling at 193 K and various stress amplitude, which were monitored by the output voltage from the piezoelectric transducer, for non-irradiated KBr single crystal.

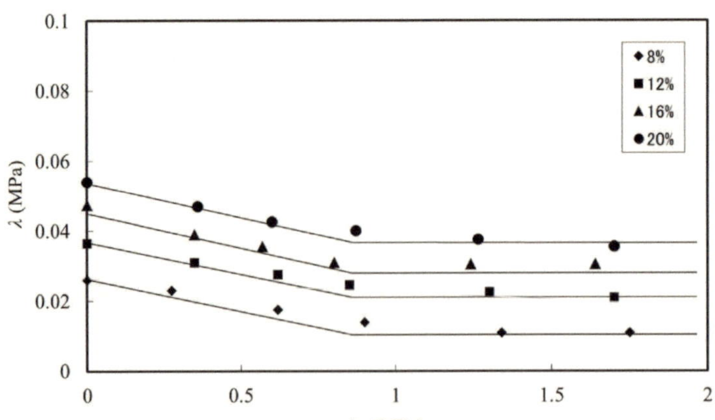

Figure 7.
*Strain-rate sensitivity (λ) versus stress decrement ($\Delta\tau$) at various strains for non-irradiated KBr single crystal at 193 K. the plotted points are obtained from **Figure 6a** and **b**.*

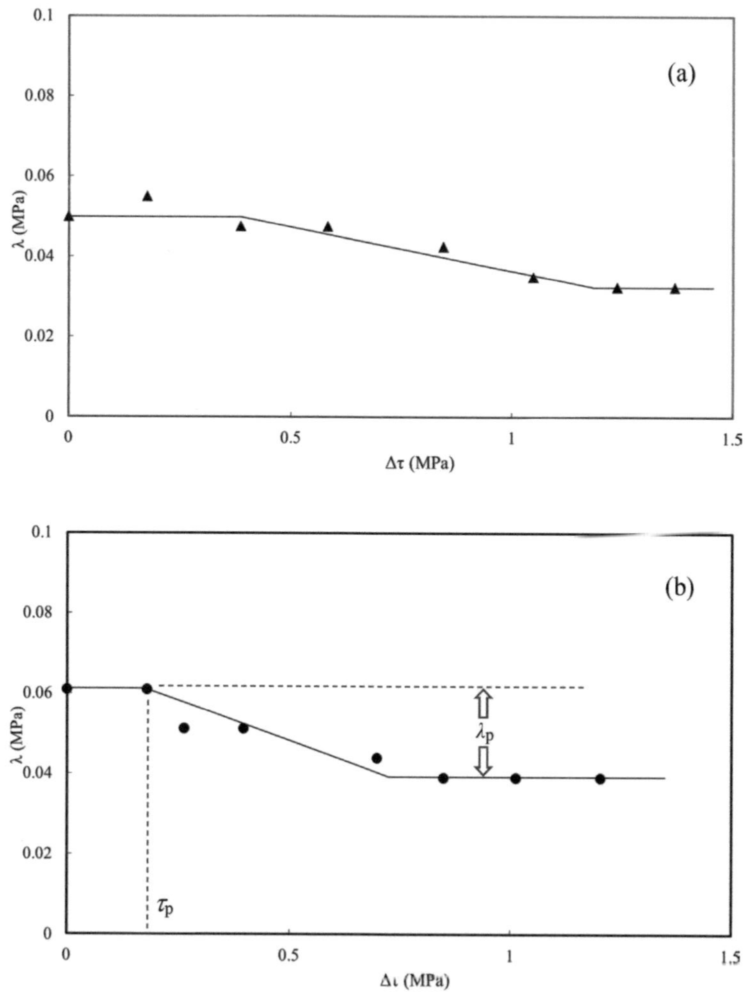

Figure 8.
Strain-rate sensitivity (λ) versus stress decrement (Δτ) at (a) 183 K and (b) 203 K for X-ray- irradiated KBr single crystals.

plateau place ranges below the first bending point within low $\Delta\tau$, and the second one extends from the second bending point within high $\Delta\tau$. The λ decreases with $\Delta\tau$ between the two bending points. The value of $\Delta\tau$ at first bending point is referred to as τ_p in **Figure 8b**. λ_p denoted in the figure is introduced later. τ_p has been considered the effective stress due to the weak obstacles such as dopants when a dislocation begins to break-away from the dopants which lie on the dislocation with the help of thermal activation during plastic deformation of alkali halide doped with monovalent or divalent ions [6–8], because τ_p depends on temperature and on type and density of the obstacle [15, 16]. Although τ_p is not observed on the relative curve of λ versus $\Delta\tau$ for the nonirradiated crystal at 193 K in **Figure 7**, τ_p becomes to appear on the relative curve at nearly 193 K by irradiating the crystals with the X-ray, as shown in **Figure 8(a)** and (**b**). Therefore, the weak obstacles are considered the X-ray-induced defects here. Namely, the appearance of τ_p in this chapter represents the effective stress due to the defects such as V_2-centres referred in the previous section (i.e., Section 3.1).

3.3 Critical temperature (T_c) and Gibbs-free energy (G_0)

Figure 9 shows the dependence of τ_p on temperature for the sample (the X-irradiated KBr crystal). τ_p tends to decrease with increasing temperature and appears to be zero at the critical temperature T_c above 500 K. **Figure 10** shows the relation between τ_p and activation volume V, which was obtained from kT/λ_p, for the thermally activated dislocation motion in the sample. k is the Boltzmann constant. λ_p is the difference between λ values at the first and second plateau places on the relative curve of λ versus $\Delta\tau$ (see **Figure 8b**) and has been regarded as the strain-rate sensitivity due to point defects [18] as expressed by the following Eq. (1).

$$\lambda_p = \frac{kT}{bl_p d} \tag{1}$$

where b is the magnitude of Burgers vector, l_p is the average spacing of point defects along dislocation, and d is the activation distance. The force-distance curve between dislocation and tetragonal defect was reported by Barnett and Nix [19]. **Figures 9** and **10** reflect the interaction between dislocation and X-ray- induced defects. The solid curves in these figures were obtained on the assumption that the force-distance profile derived by Barnett and Nix (Barnett model) is appropriate for the interaction between dislocation and the radiation-induced defect in the X-irradiated KBr crystal. Each curve, which was given by numerical calculation with the parameters (τ_{p0}, T_c, and G_0) listed in **Table 1**, agrees with the data (solid circles in these figures) analysed in terms of λ versus $\Delta\tau$ for the samples. The critical temperature T_c at which τ_p is zero is about 660 K, and the τ_{p0} value of τ_p at absolute zero is 0.67 MPa. G_0 (0.81 eV for the sample) is the Gibbs free energy for breaking away from the radiation-induced defect by dislocation at absolute zero.

The values of τ_{p0}, T_c, and G_0 are also listed for the same X-ray-irradiated NaCl, NaBr, and KCl single crystals in **Table 1**. "X-ray-irradiated" is abbreviated to "X-Irr." in the table. These values are derived on the basis of λ versus $\Delta\tau$ obtained by the above-mentioned method (i.e., the strain-rate cycling tests combined with ultrasonic

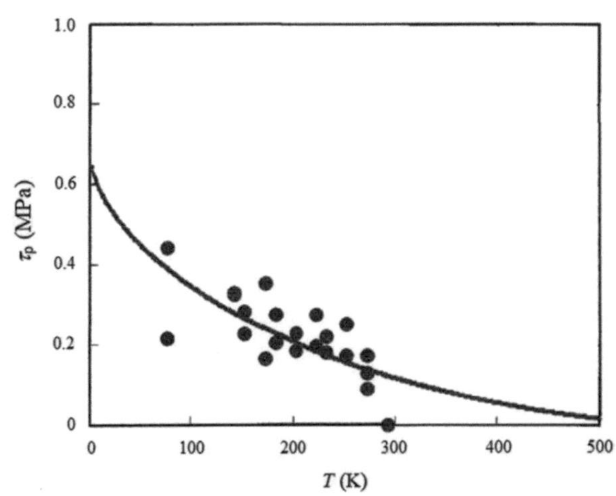

Figure 9.
Dependence of τ_p on temperature for X-ray-irradiated KBr single crystals. The solid curve is given by numerical calculation (reproduced from Ref. [17] with permission from the publisher).

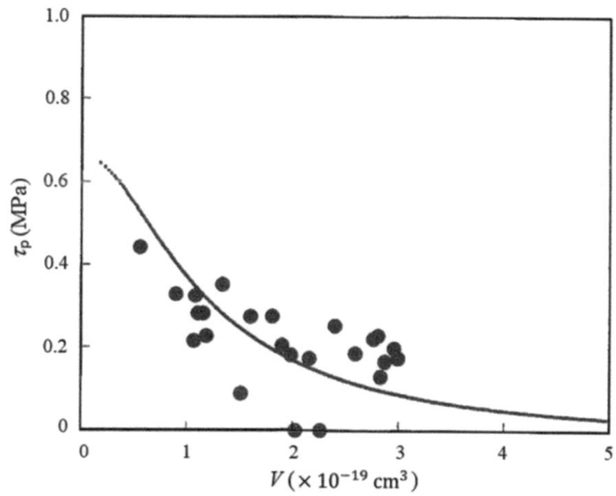

Figure 10.
Relation between τ_p and activation volume (V) for X-ray-irradiated KBr single crystals. The solid curve is given by numerical calculation (reproduced from Ref. [17] with permission from the publisher).

Crystal	τ_{p0} (MPa)	T_c (K)	G_0 (eV)
X-Irr. NaCl	1.94	346	0.39 ± 0.09
X-Irr. NaBr	1.99	369	0.76
X-Irr. KCl	1.00	528	0.87 ± 0.19
X-Irr. KBr	0.67	660	0.81

Table 1.
Values of τ_{po}, T_c, and G_o for X-ray-irradiated single crystals [17, 20].

oscillation). As for X-Irr. NaBr, the peak of V_2-centre is not clear in the absorption spectrum within the measurement [17]. In the spectrum, V_2-centre for X-Irr. NaCl and V_3-centre for X-Irr. KCl are seen as reported in the paper [20]. As given in **Table 1**, G_0 values are 0.39 ± 0.09, 0.76, and 0.87 ± 0.19 eV for X-Irr. NaCl, X-Irr. NaBr, and X-Irr. KCl, respectively. This suggests that the difference between longitudinal and transverse strains of distortion around the irradiation-induced defect, i.e. the tetragonality of defect, becomes large in the order of NaCl, NaBr, KBr, and KCl by the X-ray irradiation at room temperature.

4. Conclusion

The strain-rate cycling tests combined with ultrasonic oscillation were conducted for X-ray-irradiated KBr single crystals. Analysing the relation between λ and $\Delta\tau$ for two kinds of samples (non-irradiated and X-irradiated KBr crystals), the bending point τ_p within low $\Delta\tau$ is referred to the X-ray-induced defects in the sample. In addition, τ_p vs. T and τ_p vs. V curves reflect the interaction between dislocation and X-irradiated defects and further give the values of τ_{p0}, T_c, and G_0 for the sample. With regard to the same X-ray-irradiated NaCl, NaBr, and KCl crystals, the values of them have been obtained on the basis of τ_p vs. T or V curves by the mentioned method. G_0

values are 0.39 ± 0.09, 0.76, 0.87 ± 0.19, and 0.81 eV for X-Irr. NaCl, X-Irr. NaBr, X-Irr. KCl, and X-Irr. KBr, respectively. This suggests that the tetragonality around defect induced by the X-irradiation becomes large in the order of NaCl, NaBr, KBr, and KCl.

Acknowledgements

Dr. T. Ohgaku, as well as S. Matsumoto are acknowledged for his collaboration in the analysis on λ and $\Delta\tau$ data, as well as for his experimental assistance.

Conflict of interest

The author declares no conflict of interest.

Author details

Yohichi Kohzuki
Department of Mechanical Engineering, Saitama Institute of Technology, Fukaya, Japan

*Address all correspondence to: kohzuki@sit.ac.jp

References

[1] Sibley WA, Sonder E. Hardening of KCl by electron and gamma irradiation. Journal of Applied Physics. 1963;**34**: 2366-2370. DOI: 10.1063/1.1702747

[2] Nadeau JS. Radiation hardening in alkali-halide crystals. Journal of Applied Physics. 1964;**35**:1248-1255. DOI: 10.1063/1.1713603

[3] Reppich B. Radiation hardening of alkali halides and magnesium oxide. Scripta Metallurgica. 1971;**5**:289-293. DOI: 10.1016/0036-9748(71)90197-9

[4] Tanimura K, Fujiwara M, Okada T, Hagihara T. Mechanism of low-temperature radiation hardening in alkali halides. I. Single halogen interstitials as the hardening agent at low temperature. Journal of Applied Physics. 1978;**49**: 5452-5456. DOI: 10.1063/1.324513

[5] Meshii M. Interaction of glide dislocations with interstitials and lattice vacancies. In: Eyre BL, editor. Proceedings of Symposium on the Interactions between Dislocations and Point Defects. Atomic Energy Research Establishment, Harwell; 1968. pp. 566-603

[6] Ohgaku T, Matsunaga T. Interaction between dislocation and divalent impurity in KBr single crystals. IOP Conference Series: Materials Science and Engineering. 2009;**3**:012021. DOI: 10.1088/1757-899X/3/1/012021

[7] Kohzuki Y. Study on the interaction between a dislocation and impurities in KCl:Sr single crystals by the Blaha effect-Part IV influence of heat treatment on dislocation density. Journal of Materials Science. 2009;**44**:379-384. DOI: 10.1007/s10853-008-3150-8

[8] Kohzuki Y. Bending angle of dislocation pinned by an obstacle and the Friedel relation. Philosophical Magazine. 2010;**90**:2273-2287. DOI: 10.1080/14786431003636089

[9] Sirdeshmukh DB, Sirdeshmukh L, Subhadra KG. Alkali Halides. Berlin Heidelberg: Springer-Verlag; 2001. p. 41. DOI: 10.1007/978-3-662-04341-7

[10] Urusovskaya AA, Petchenko AM, Mozgovoi VI. The influence of strain rate on stress relaxation. Physica Status Solidi A. 1991;**125**:155-160. DOI: 10.1002/pssa.2211250112

[11] Zakrevskii VA, Shul'diner AV. Dislocation interaction with radiation defects in alkali-halide crystals. Physics of the Solid State. 2000;**42**:270-273. DOI: 10.1134/1.1131195

[12] Sirdeshmukh DB, Sirdeshmukh L, Subhadra KG. Micro- and Macro-Properties of Solids. Berlin Heidelberg: Springer-Verlag; 2006. p. 176. DOI: 10.1007/3-540-31786-4

[13] Seitz F. Color centers in alkali halide crystals. Reviews of Modern Physics. 1946;**18**:384-408. DOI: 10.1103/RevModPhys.18.384

[14] Seitz F. Color centers in alkali halide crystals II. Reviews of Modern Physics. 1954;**26**:7-94. DOI: 10.1103/RevModPhys.26.7

[15] Ohgaku T, Takeuchi N. Interaction between a dislocation and monovalent impurities in KCl single crystals. Physica Status Solidi A. 1992;**134**:397-404. DOI: 10.1002/pssa.2211340210

[16] Kohzuki Y, Ohgaku T, Takeuchi N. Interaction between a dislocation and impurities in KCl single crystals. Journal of Materials Science. 1993;**28**:3612-3616. DOI: 10.1007/BF01159844

[17] Kohzuki Y, Ohgaku T. X-ray-induced defects as obstacles to dislocation motion in alkali halide crystals. Journal of Materials Science. 2017;**52**:3959-3966. DOI: 10.1007/s10853-016-0657-2

[18] Kohzuki Y, Ohgaku T, Takeuchi N. Interaction between a dislocation and various divalent impurities in KCl single crystals. Journal of Materials Science. 1995;**30**:101-104. DOI: 10.1007/BF00352137

[19] Barnett DM, Nix WD. The interaction force between tetragonal defects and screw dislocations in cubic crystals. Acta Metallurgica. 1973;**21**:1157-1168. DOI: 10.1016/0001-6160(73)90032-1

[20] Ohgaku T, Migiuuma S, Nagahira D. Interaction between dislocation and defects induced by X-irradiation in alkali halide crystals. Radiation Measurements. 2011;**46**:1385-1388. DOI: 10.1016/j.radmeas.2011.05.067

Section 3

Organic Rocks

Chapter 9

Mineralogical Properties of Moroccan Rif Bituminous Rocks

Khalihena Groune A.L.

Abstract

Bituminous rocks, including oil shales, are sedimentary rocks. They include a group of rocks like shales, marls and carbonates. Containing 60–90% mineral matter and less organic matter. Five techniques namely, XRD, FRX, ICP/AES, MEB-EDX and FTIR analysis were used to characterize the the mineral material of Moroccan Rif bituminous rock. The results obtained by these methods showed that the dominant mineral phase in all samples is quartz (70.04–84.46 wt%). The TA sample contains high amounts of calcium oxide (CaO) and magnesium oxide (MgO), as well as a task amount of carbonates compared to the other samples. These quantities were confirmed by ICP/AES, MEB-EDX and ETIR. According the XRD analysis, the illite and the chlorite (clay minerals) and other mineral are existed in various proportions of low weight.

Keywords: bituminous rocks, spectroscopy, XRD analysis, upper cretaceous, Moroccan Rif

1. Introduction

Morocco is almost entirely dependent on oil and gas imports to cover its energy needs. However, large oil shale deposits are available in many parts of the country but remain untapped. According to the latest study recently published by the World Council of Energy-on-energy reserves in the world, Morocco was ranked in sixth position, with respect to the world's reserves of oil shale. The Rif deposit (North of Morocco) is one of three major Moroccan deposits which come up to ground level. It was discovered during the thirties in the last century and has been the least studied in comparison with the other deposits of Morocco (Al-Tamahdt and Tarfiyah), and the reason for this is due to the low amount of organic matter [1, 2].

Bituminous rocks, including oil shales, are sedimentary rocks. They include a group of rocks like shales, marls and carbonates. Containing 60–90% mineral matter and less organic matter [3]. Knowledge of minerals and the distribution of elements in this type of rock is very important for developing various exploitation plans.

This work focuses on the Moroccan Rif deposit, which is the first deposit in Morocco of bituminous rocks. Which was discovered by chance by extracting construction materials [4–6]. According to the literature, this region has not been studied in detail. The inorganic composition was investigated using several techniques: FTIR analysis, X-ray diffraction analysis (XRD), X-ray fluorescence spectrometry (XRF), inductively coupled plasma atomic emission spectroscopy (ICP-AES) and scanning electron microscopy (SEM), to characterize the Moroccan Rif bituminous rock samples.

2. Geological setting of Moroccan Rif

The Moroccan Rif is a western end of the Maghreb chains which is a segment of a larger whole i.e.; the Alps [7]. The Moroccan Rif includes three major geological domains or nappe complexes: i) The Internal Zone, or Alboran Domain, ii) The Flysch nappes and iii) The External Zone (**Figure 1**) [7].

The External Zone mainly results to the separation in the Miocene of the Mesozoic sedimentary cover and of Paleogene deposited on northern margin of the plate in Africa [8–10]. It consists by three types of units which are divided from NE to SW and top to bottom, into the Intrarif, Mesorif and Prerif units [8].

The Flysch nappes generally overlie the External Rif with the exception of some klippes that are located on the Internal Zone as a result of a complex evolution [11]. These Flysch units are of Cretaceous-Early Miocene age. Their basement does not crop out at present and may be considered as the deposits of a swell located on ocean or transitional crust in between the External and Internal Zones [8, 9, 12].

The Internal Zone consists of old socle of Paleozoic nappes and of limestone chain. These elements are derived from the dilaceration of the Alboran plate. The internal zone outcrops along the Mediterranean coast (Alboran Sea) between Ceuta and Jebha where it disappear under the sea thanks to a big transfer fault [13] to reappear later in the west of Al Hoseima (**Figure 1**).

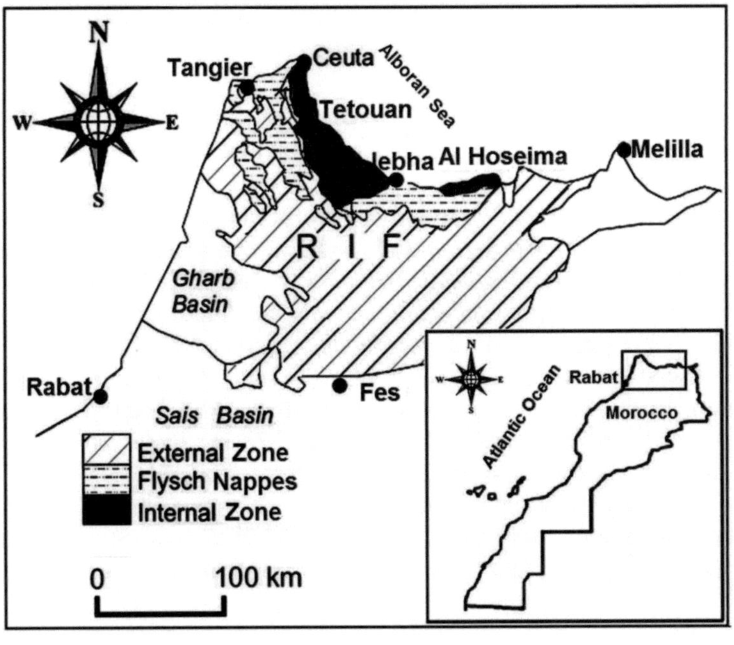

Figure 1.
Geological map of the Moroccan Rif.

3. Sampling and stratigraphy of study areas

The studied samples of Moroccan Rif bituminous rocks are located in four sites called: Tangier, Tetouan, Bab Taza and Arba Ayacha (**Figure 2**). A sample of 20 kg has been collected from each station (20–80 cm depth) and only one sample has been collected from each site.

TA sample of Tangier site is from Upper Cretaceous – bituminous marls intercalated with limestone in the unity of the Melloussa-Chouamat (from Massylienne nappe of Flyschs nappes) (**Figures 3A** and **4a**) [16].

TE, BT samples of Tetouan and Bab Taza sites respectively are from Upper Cretaceous bituminous Phtanites include in schistose marls in the unity of Internal Tangier and unity of Ketama – External Tangier successively (from Intrarif of External Zone) (**Figure 3B, C** and **4b, c**) [14, 15].

AA sample of Arba Ayacha site is from Upper Cretaceous oil shale intercalated with limestone and white marls in the unity of the Melloussa-Chouamat (from Massylienne nappe of Flyschs nappes) (**Figures 3A** and **4d**) [16, 17].

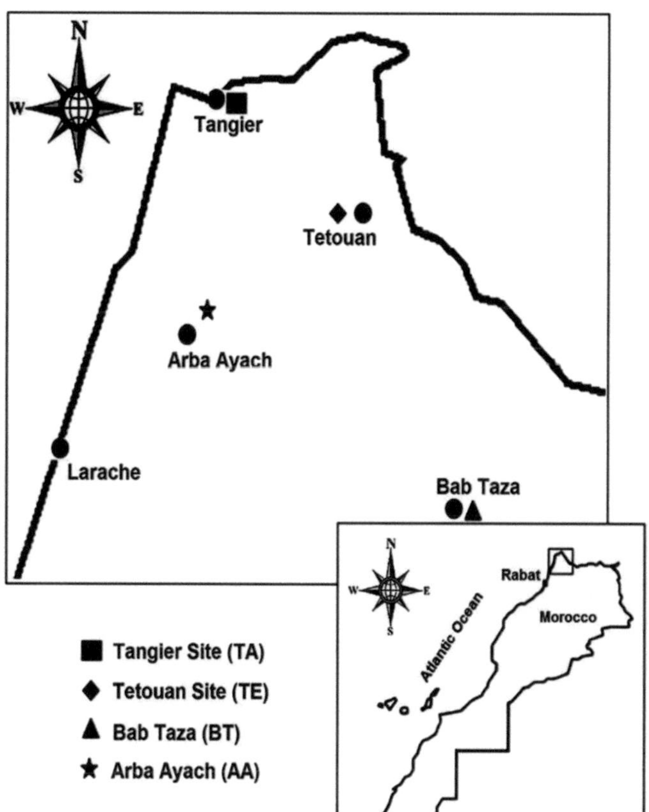

Figure 2.
Study area showing the bituminous rocks sampling location.

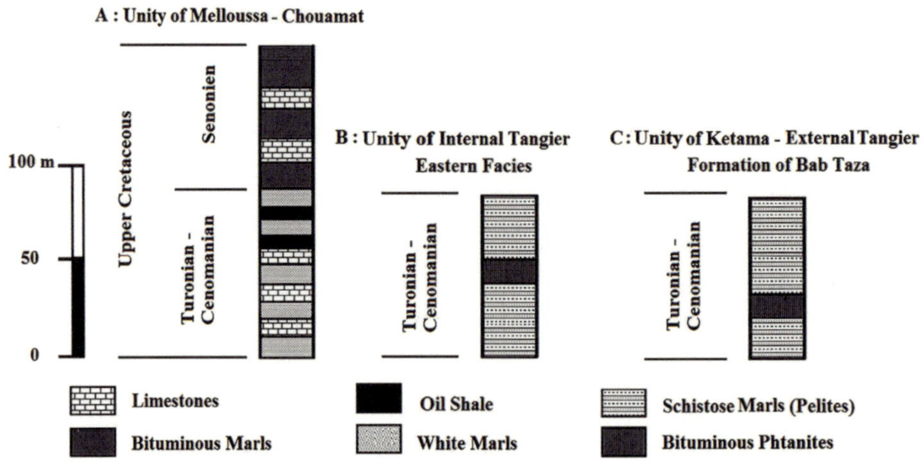

Figure 3.
Stratigraphic columns of upper cretaceous age of the unity of Melloussa-Chouamat (Flyschs nappes) and the Intrarif of the unity of Tangier (external zone), showing the investigated sample position (A: [14] (B: [14]) (C: [15]).

Figure 4.
Photographs for the four sites of the samples.

4. Materials and methods

The four samples (TA, TE, BT, AA) were subsequently dried, crushed, homogenized and sieved to <180 μm in order to undertake the following analytical operations (**Figure 5**).

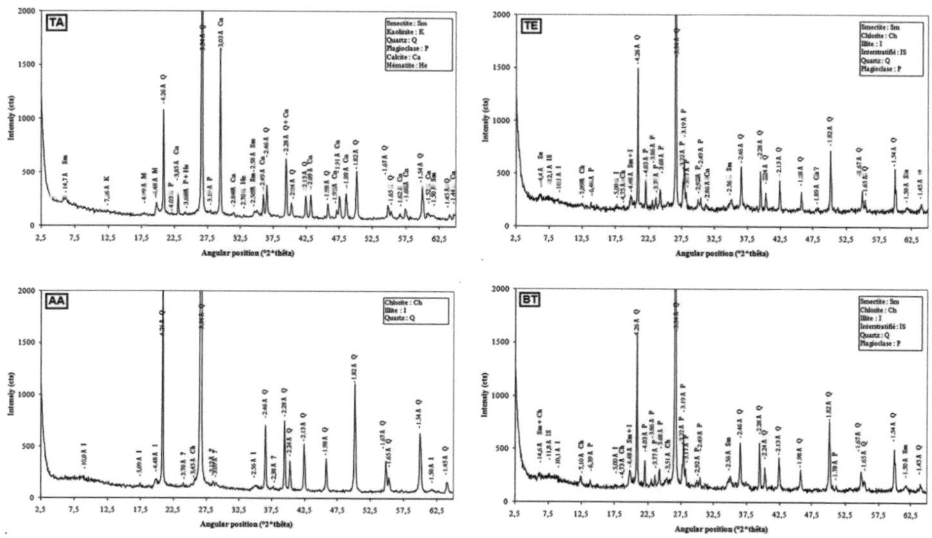

Figure 5.
Diffractogram of main mineral constituents of crude samples.

4.1 X-ray diffraction analysis

The crude Samples were analyzed by X-ray diffraction to determine the main mineral constituents. The analyses were carried out on a Brucker D8 Advance diffractometer equipped with a copper anticathode ($\lambda CuK\alpha$ = 1.541838 Å), the analytical conditions were 40 kV, 40 mA, Ni filter, angular range 2θ° [2.5 to 65].

4.2 X-ray fluorescence spectrometry (XRF)

The XRS analytical technique are used to measure and identify the inorganic elements concentration in the sedimentary rocks. The rock raw samples were analyzed using an FX AXIOS PW4400 X-ray fluorescence (XRF) spectro-meter system.

4.3 Inductively coupled plasma-atomic emission spectrometry (ICP-AES)

The ICP-AES analytical technique are used to determination of mass concentrations of inorganic elements in the sedimentary rocks [18, 19]. The rock raw samples were analyzed using a Jobin Yvon Ultima 2 m spectrometer after digestion in HF acid.

4.4 Scanning electron microscopy (SEM)

The SEM is an important qualitative or semi-quantitative technique for knowing particle composition and morphology. As well as for analyzing surface elements, SEM is usually used to obtain topographical images of mineral grains and to determine their distribution.

Each sample of the whole rock was analyzed by a Cambridge Stereoscan 120 instrument coupled with an elemental analysis (EDX) unit. The apparatus was adjusted at 7–10 mm WD, 30 kV HV and 10,000X magnification.

4.5 FTIR analysis of crude sample

FTIR spectra were recorded on KBr discs. Each crude sample (~0.5 mg) was added to 100 mg of dry KBr powder, uniformly mixed and reground. The entire sample was transferred to a die and pressed under vacuum in the standard way. Spectra were recorded between 400 and 4000 cm^{-1} using a Vertex 70 Model spectrometer.

5. Results and discussion

5.1 X-ray diffraction analysis

The powder diffractograms of the crude samples (**Figure 5**) show that the major constituents of these samples are: quartz, calcite and plagioclase with predominance of quartz, with some traces of hematite only in TA sample. The plagioclase is present in all samples except the AA sample [20].

Clay minerals are rare or at least less visible on untreated material (no cementite dissolution). Semi-quantification of major minerals using their major reflections is provided in **Table 1** [20].

The mineral composition of the fraction smaller than 2 μm of the samples studied is qualitatively shown in **Table 2**. After decomposing cements (carbonate and organic matter), it appears that the most abundant clay minerals are not chlorite or illite, but phases of a swelling type smectite or interstratified illite/smectite. Smectite dominates in the sample TA while for the samples BT and TE it is interstratified. Furthermore, it is noticed that the AA material is virtually devoid of clay fraction, except some traces of chlorite and illite on the threshold of detection limit.

Sample	Quartz	Calcite	Plagioclase	Hematite	Clay
TA	++++	+++	tr	tr	++
TE	++++		++		+
BT	++++		++		++
AA	+++++				+

Table 1.
Semi-quantification of major minerals in crude samples.

Sample	Chlorite	Chlorite	Chlorite	Chlorite	Chlorite
TA	+	+++	+	+	
TE	++		+++	+	+
BT	++		+++	+	+
AA	tr			tr	

Table 2.
The mineral composition of the <2 μm fraction of the samples.

5.2 X-ray fluorescence spectrometry (XRF)

The XRF analysis of the samples (**Table 3**) shows very high of their content of quartz (70.04–84.46 wt%). The mass concentrations of calcium oxide (CaO) and magnesium oxide (MgO) are respectively 0.013–0.84 wt% and 0.47–1.32 wt%, indicating very low content of dolomite and calcite (carbonates) in the samples. The mass concentration of aluminum oxide (Al_2O_3) ranges between 3.90 and 12.34 wt%, ferric oxide (Fe_2O_3) between 0.74 and 3.56 wt% and sulfur trioxide (SO_3) from 0.08 to 0.74 wt%; the two last proportion ranges indicate the low content of pyrite (FeS_2).

5.3 ICP-AES analyses

Table 4 shows the mass concentrations of inorganic elements. Silica is between 32.2–39.3% by weight, followed by aluminum between 2.44–5.88% by weight.

	TA	TE	BT	AA
SiO_2	80.22	70.04	73.78	84.46
Al_2O_3	7.929	12.34	11.87	3.908
Fe_2O_3	2.172	3.569	1.808	0.644
CaO	0.84	0.36	0.191	0.013
MgO	1.216	1.31	1.323	0.48
TiO_2	0.2	0.3	0.31	0.09
SO_3	0.1	0.11	0.08	0.74
K_2O	0.64	0.73	0.92	0.23
ZnO	0.14	0.05	0.05	0.05
Na_2O	0.31	1.815	1.246	0.18
MnO_2	0.034	0.09	0.04	0.02
CuO	0.03	0.02	0.04	0.01
P_2O_5	0.05	0.08	0.1	0.02
V_2O_5	0.024	0.1	0.06	0.1
Cr_2O_3	0.02	—	0.04	0.02
BaO	0.02	0.02	0.02	—
NiO	0.01	0.02	0.02	0.01
SrO	0.01	0.031	0.004	0.01
Rb	0.01	0.004	0.01	—
CoO	—	0.04	0.01	0.01
As_2O_3	—	0.031	—	0.004
PbO	—	—	0.01	—
Y_2O_3	—	—	—	0.001
L.O.I	6.01	9.013	8.08	8.983
Total	99.991	99.995	100.012	99.99

Table 3.
Major, minor and trace element XRF analyses of 4 samples, wt%.

Sample, wt%	Al	Ca	Fe	K	Mg	Na	P	S	Si	Ti
AA	2.44	0.032	0.228	0.563	0.229	0.170	0.012	0.304	39.300	0.031
BT	5.88	0.159	0.596	0.996	0.350	1.080	0.013	0.0209	34.100	0.148
TE	5.34	0.251	1.208	0.888	0.350	0.860	0.035	0.0209	32.900	0.076
TA	3.31	5.607	1.045	0.640	0.498	0.203	0.053	0.04	32.200	0.074

Table 4.
Mass concentrations of inorganic elements in the bituminous rock samples of 4 samples by ICP-AES analysis.

The calcium mass concentration is much higher (5.6 wt%) in the TA sample than the other samples. Mass concentrations of the rest of the inorganic elements are equal to or less than 1% by weight. This was confirmed by X-ray spectrometry.

5.4 Scanning electron microscopy (SEM)

According to SEM observations, there are some differences in particle shape between samples. **Figure 6** shows the SEM–EDX analysis of the studied rock samples. It is seen that the surface of the TA sample is mainly composed of small particles on the order of 1 μm. While the surface of the TE sample appears compact and is dominated by particle agglomerations. For sample BT, the microstructure consists of large holes and heterogeneous particles of different sizes, while the surface of sample AA consists of homogeneously dispersed particles.

The EDX analysis shows that the surface of the samples is dominated by quartz and aluminum oxide species, especially for the TE and BT samples. The carbon peaks observed in the EDX spectrum (**Figure 6**) indicate the total carbon content, which includes organic carbon (as organic matter) and metallic carbon (as carbonate). These results are in good agreement with the results obtained with the previously mentioned analytical techniques.

5.5 FTIR analysis of crude sample

FTIR spectra of the crude samples are shown in (**Figure 7**). Based on the known bituminous rocks data [21], the absorption bands observed in the spectra of the four studied samples were attributed as follow:

Characteristic bands of organic matter: The bands within the vibration range 400–875 cm^{-1} belong to aromatic C-H structure, alkyl-Si and sulfides ... The absorptions between 1030 and 1170 cm^{-1} corresponding to CC, CO and CN stretching and HCC, CCC bending bands. The vibration bands range 1500–1900 cm^{-1} belong to aromatic CH=CH structure and COO^-, CO of naphthalene groups. The bands appear between 2300 and 2500 cm^{-1} belong to C=C and C=N of unsaturated liaisons. The bands between 2800 and 3000 cm^{-1} correspond to stretching of aliphatic chains. The band within the vibration range 3000–3700 cm^{-1} belongs to NH and OH stretching of (alcoholic, phenolic and carboxylic ...) groups and also of aromatic C-H structure.

Characteristic bands of mineral matter: In spectra of crude rocks, the absorption bands of inorganic compounds are the most intense and often mask those of

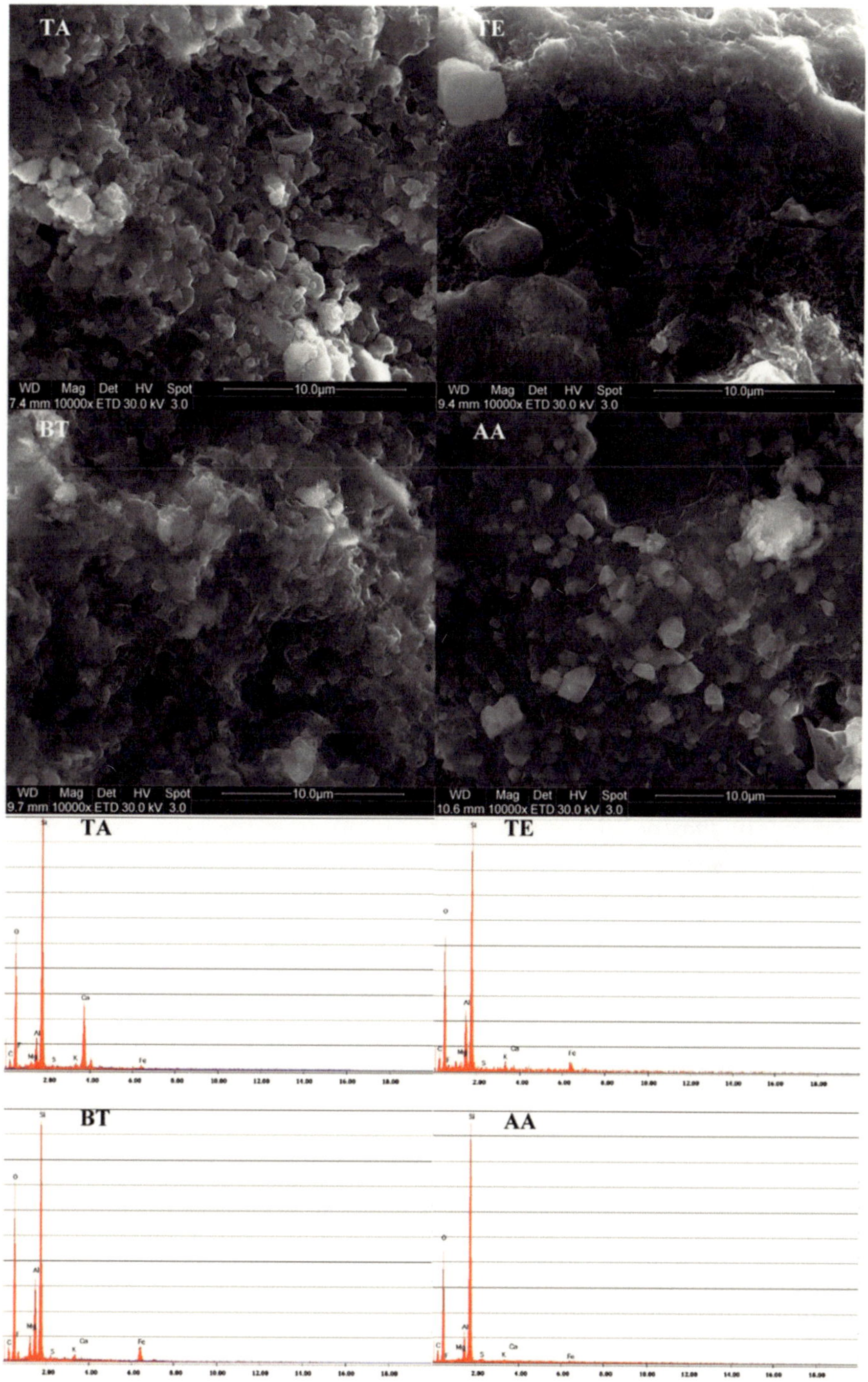

Figure 6.
SEMs of raw bituminous rocks samples: A) for TA sample, B) for TE sample, C) for BT sample, D) for AA sample. Taken from [20].

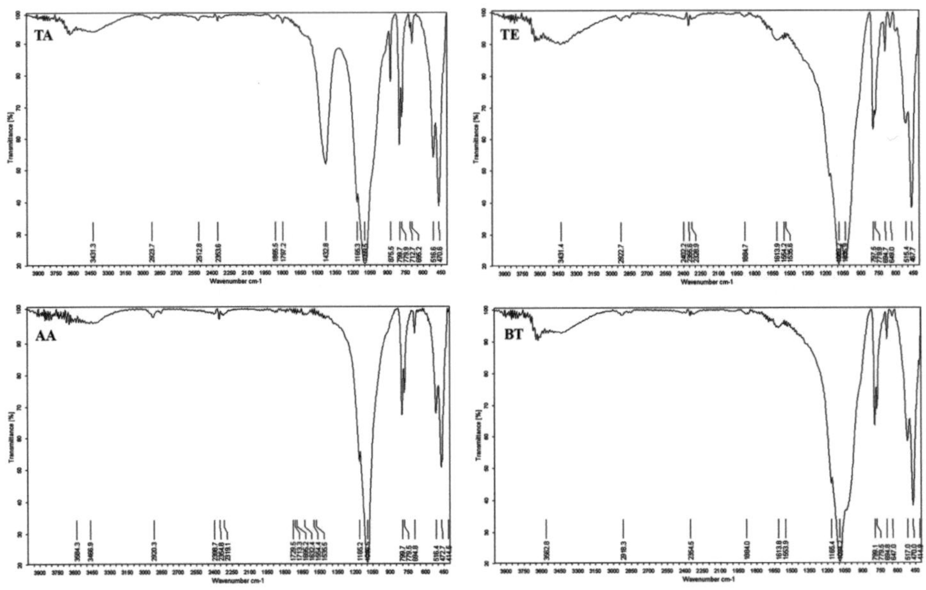

Figure 7.
FTIR spectra of the crude samples.

fundamental bands of organic compounds. The Fe_2O_3 phase may be characterized by absorption bands at 520, 460 and 470 cm^{-1} which are difficult to distinguish from those of the organic matter bands. The 1433 cm^{-1} absorption band visible only in the spectrum of TA sample belongs to the bending vibration of carbonate ($CaCO_3$) and confirms our findings of XRD results. The bands characterizing clays correspond to frequencies between 3000 and 3700 cm^{-1} in the OH region (H_2O).

The bands due to vibration of Si-O-M structure fall into three series. The first set is displayed with very intense frequencies to 1035, 1094, 1096, 1099, 1089 and 1165 cm^{-1}. The second set is less intense and the bands are located at frequencies between 600 and 800 cm^{-1}. As for the last of medium intensity, it is between the frequencies 460 and 520 cm^{-1}. The predominance of quartz and element of silica (Si) in the mineralogical composition of the four examined samples have already been proved by XRD analysis.

6. Conclusion

The study of bituminous rocks of Moroccan Rif provides important information about representative samples in this deposit. Mineralogical composition of the crude samples is dominated by the quartz with the presence of calcite (carbonate) only in the TA sample (bituminous marls) and also confirmed by FTIR analysis. The clay minerals and other mineralogical components are present as different low weight proportions.

The mineralogical components that were found by X-ray fluorescence spectrometry (XRF) and X-ray diffraction (WRD), show that the samples of bituminous rocks studied in the Moroccan Rif, characterized by a high content of silica (quartz), and lower contents of dolomite, calcite (Carbonate), and Pyrite. That was confirmed by other analyzes: ICP/AES, MEB-EDX.

Acknowledgements

The author is grateful for the support for this work from the University of Mohammed V - Rabat, Morocco.

Author details

Khalihena Groune A.L.
Sciences and Technologies Faculty, Chemistry Department, University of Nouakchott, Nouakchott, Mauritania

*Address all correspondence to: khalihenagroune@gmail.com

References

[1] Halim M. Etude de la matiére organique sédimentaire, Gisements de Timahdit et Tarfaya [Thèse de Doctorat Es-Sciences]. Rabat, Maroc: Université Mohammed V; 1993

[2] Amblès A, Halim M, Jacquesy JC, Vitorovic D, Ziyad M. Characterization of kerogen from Timahdit shale (Y-layer) based on multistage alkaline permanganate degradation. Fuel. 1994;**73**:17-24

[3] Yen TF, Chilingarian GV. Introduction to Oil Shales. Amsterdam: Elsevier; 1976. pp. 1-12

[4] Berthelot C. Carbonization of Oil Shales in France. France: Factory St. Hilaire (Allier); 1938

[5] Berthelot Ch. The deposits of oil shales and limestones of Tangier and Mogador. 1938

[6] Nejma M. Contribution in the Development from Rock to Kerogen of Timahdit. Rabat: Mohammed V University; 1989

[7] Durand-Delga M, Fontboté JM. Le cadre structural de la Méditerranée occidentale. In: Géologie des chaines alpines issues de la Téthys. In 26ème th International geological Congress. Paris: Mém. B.R.G.M; 1980. pp. 67-85

[8] Suter G. Carte géologique du Maroc: Carte structurale de la chaîne rifaine – Echelle 1/500 000. Notes et Mém. Serv. Géol. Maroc, Rabat. 1980;**245**

[9] Wildi W. La chaine tello-rifaine (Algérie, Maroc, Tunisie) : structure, stratigraphie et évolution du Trias au Miocène. Géol. Dyn. Géorgr. Pys. 1983;**21**:201-297

[10] Ben YA. Évolution Tectono-sédimentaire du Rif externe centro-occidental (région de Msila et Ouazzane, Maroc). La marge affricaine du Jurassique au Crétacé, les bassins néogènes d'avant-fose [Thèse de Doctorat d'Etat]. France: Université de Pau et Pays de l'Adour; 1991

[11] Lespinasse P. Géologie des zones externs et des Flyschs entre Chaouen et Zoumi (Centre de la chaine rifaine du Maroc) [Thèse Doc. D'Etat és-Sciences]. Paris: Univ. Toulouse; 1975. p. 284

[12] Durand-Delga M, Rossi P, Olivier P, Puglisi D. Situation structurale et nature ophiolitique des roches basiques jurassiques associées aux flyschs maghrébins du Rif (Maroc) et de Sicile (Italie). Comptes Rendus Geoscience. 2000;**331**:29-38

[13] Benmakhlouf M, Galindo-Zaldívar J, Chalouan A, et al. Inversion of transfer faults: The Jebha-Chrafate fault (Rif, Morocco). Journal of African Earth Sciences. 2012;**73-74**:33-43

[14] Durand-Delga M, Kornprost J. Carte géologique du Tétouan – Ras Mazari 1/50.000. Notes et Mém. Serv. Géol. Maroc. 1985;**292**

[15] Kornprost J, Wildi W, Nold M, Gutnie M, Lespinasse P. Carte géologique du Rif: Bab Taza 1/50.000. Notes et Mém. Serv. Géol. Maroc. 1980. 288

[16] Durand-Delga M, Olivier P. Carte géologique du Rif Méloussa 1/50.000. Notes et Mém. Serv. Géol. Maroc. 1988;**296**

[17] Saadi M, Bouhaouli A, Hilali EA, Ider EH, Ralhali T. Les roches bitumineuses marocaines, Numéro.

Spécial Consacré aux schistes bitumineux au Maroc, Géologie et Énergie. Ministère de l'Énergie et des Mines. 1981;**50**

[18] Wilkerson CL. Trace metal composition of Green River retorted shale oil. Fuel. 1982;**61**(1):63-70

[19] Orupõld K, Habicht J, Tenno T. Leaching behaviour of oil shale semicoke: Compliance with the waste acceptance criteria for landfills. Oil Shale. 2008;**25**(2):267-275

[20] Groune K, Halim M, Arsalane S. Thermal and mineralogical studies of Moroccan Rif bituminous rocks. Oil Shale. 2013;**30**(4):536-549

[21] Abdelfatah A. Etude spectroscopique d'une série d'alcanes ramifies, des schistes bitumineux, de leur évolution thermique et de leurs extraits organiques. Thèse de Doctorat Es-Science physique, Université Mohammed V, Rabat, Maroc. 1999